BASKET CASE

What's happening to Ireland's food

PHILIP BOUCHER-HAYES & SUZANNE CAMPBELL ∾

Gill & Macmillan

Gill & Macmillan Ltd
Hume Avenue, Park West, Dublin 12
with associated companies throughout the world
www.gillmacmillan.ie

978 07171 4579 9

Index compiled by Cover to Cover
Print origination by O'K Graphic Design, Dublin
Printed by Scandbook AB, Sweden

This book is typeset in 12/14.5 Minion.

The paper used in this book comes from the wood pulp
of managed forests. For every tree felled, at least one tree
is planted, thereby renewing natural resources.

A CIP catalogue record for this book is available from the
British Library.

5 4 3 2

CONTENTS

ACKNOWLEDGEMENTS   VII

INTRODUCTION   IX

PART ONE

1. What's that terrible smell? — 2
2. Who do you think you are? — 16
3. The way we ate — 34
4. Who is rural Ireland? — 53
5. The urban playground — 67
6. The land of heart's desire — 79
7. Consuming ourselves — 89
8. Green is the new black — 104

PART TWO

9. Sucking diesel on the road to Dublin — 120
10. Every little helps — 128
11. Sense and sensibility — 139
12. Square pegs into round holes — 151
13. Mythbusting: A tale of two markets — 164
14. Where does it all come from? — 174
15. Chickens with page three breasts — 188
16. The rise and rise of the superbug — 197
17. They don't go away, you know — 211
18. Could food and farming save us all? — 221

CONCLUSION
Too lazy to read it all? Have a look at this
  instead   231

NOTES   238
INDEX   245

ACKNOWLEDGMENTS

There are countless farmers, horticulturalists, fishermen and food producers the length and breadth of the island who have informed this book. There are too many to name and unfortunately many have insisted that their contribution remain anonymous. This says as much about the unfair relationship between farmers and supermarkets as anything else you will read in this book.

We received sound advice and colourful input from Agriculture Minister turned bookie turned talk radio host Ivan Yates, and from Suzanne's former presenter on *Ear to the Ground* and pin-up girl to the farming community, Mairead McGuinness, MEP. Paddy Wall, Professor of Public Health in UCD, was, as always, a helpful pillar of common sense. Outgoing president of the IFA, Padraig Walshe, and the original of the foodie species, Darina Allen, gave of their time and expertise. *Ear to the Ground*'s farmer/reporters, Ronan Clarke, David Kavanagh and Darragh McCullough, and fellow producer/directors are also owed a debt of thanks. So are the countless farm families that opened up their homes and told their stories throughout Ireland, Europe and Asia. Also generous with their time were Tara Buckley of RGDATA, Philip's colleagues in RTÉ Radio Agriculture, Damien O'Reilly and Frances Shanahan, and Sean McConnell of *The Irish Times*, who had his brains picked even when he didn't know it.

We owe a debt of gratitude to the researchers and academics in agencies such as Teagasc, Bord Bia, The Food Safety Authority of Ireland and Agri Aware. Their work has been quoted throughout. Most supermarket chains operating in Ireland do so from behind an unhelpful wall of silence or indifference to genuine inquiry, which is why Donald McKay, the managing director of Aldi Ireland, is owed a special debt of thanks for being so transparent.

Restaurateurs Kevin Thornton, Ross Lewis, Teresa Carr, Kelvin Rynhart and Richard Corrigan helped us get to grips with whether the food craze of the boom years had actually made Irish people any more knowledgeable about food or not. Restaurant reviewer Tom Doorley, Trevor White, author of *Kitchen Con*, and Ross Golden Bannon, editor of *Food and Wine*, should also be thanked for the same reason.

We would like to thank our families and friends, particularly Brian O'Brien, Paul McGovern, Suzanne Meagher and Francesca Walsh, for their support. At Gill & Macmillan, we would like to thank Sarah Liddy, Aoife O'Kelly, Kristin Jensen and Jane Rogers for helping to turn our manuscript into this book, which we hope will be a stimulus for much-needed and better-informed debate about the future of what we eat and who we are.

INTRODUCTION

In the closing months of 2008, a friend was leaving a Temple Bar pub in Dublin on a Saturday evening. Making his way on to the street after a few pleasant pints, he was forced to stop and wait while another patron pushed a finger against one nostril and hoovered up lines of coke from one of the granite steps. Unlike most coke users in Ireland, this man's desperate need for the stuff had obviously become so urgent that he hadn't had time to search for a nearby toilet.

It all felt uncomfortably like a scene from the dying days of the Third Reich, when the Nazi authorities, knowing disaster was at hand, lost all inhibitions, welcomed in the prostitutes and drank themselves to death before worse could befall them. When societies experience a huge sense of flux, or disaster, people begin to behave strangely. Bacchanalian activities previously hidden behind closed doors burst out into the middle of streets. Reticence or embarrassment go out the window. Perhaps gentle Dublin was finally having its Berlin moment.

Earlier that week, our government met at the midnight hour to shore up two of Ireland's oldest banks. The world's financial markets were in freefall and generally things weren't looking so good. It's not surprising that some people began to feel the cold hand of panic grip their insides. What followed was the realisation that most of our economic success in Ireland was an extremely fragile stack of cards that had already begun to fall: we just didn't know it yet. For several years, rising property prices had made us all feel rich. And just like cocaine, credit was cheap and easy, and once you started it was very hard to give the bloody stuff up. Luxury cars flew out of showrooms and shopping in New York became commonplace. We had become transformed from a bunch of culchies to a nation that considered itself urbane, well-travelled and wealthy.

What was so surprising about the boom years in Ireland was how quickly we became hell-bent on leaving our origins behind us and reimagining ourselves as something else. What had defined previous generations—the struggle for land, opportunity, employment and political freedom—were ideas that were abandoned so quickly that the door was still swinging behind us. In particular, we behaved as though we had never had a rural, agricultural past, had never worn hand-me-down clothes, had anything mended or known any practical skills. Farming became a dirty word. We looked to America and the UK for our popular culture icons (celebrities) and to Europe for our holiday homes and olive tapenade. Where Ireland was fashionable it was a new Ireland; a kind of West London meets Nantucket in the fish restaurants of Schull or around cutesy painted tableware at Avoca Handweavers.

Food and eating were one of the ways in which we were desperate to redefine ourselves. The spud went out the window. In came prosciutto and sushi. Irish cooking and Irish chefs flourished, but the origin of our food, the most traditional of industries in Ireland—farming—became something we pretended to know nothing about. Our connection with growing food, and our knowledge of food, had completely broken down. We failed to relate what was on our plate to how we lived.

Despite eating out more and sampling foodstuffs from around the world, did we know more about food at the end of the boom years? Or did we actually know less? We were more time-strapped, so we shopped more in supermarkets. We bought our food from petrol stations, newsagents and fast food chains. Food was quick and cheap and so was our attitude to it. It was only when oil and commodity prices rocketed in 2008 that people began to examine what they spent on food. We had let our connection to food become distant: how it was grown and processed, what was in it and the price at which it was sold was now defined by multinationals, for whom people were units of consumption. In Ireland at the end of 2008, people didn't grow food on the land, they did lines of coke off it.

Our traditional love of land had changed from valuing it as a resource for survival to wanting it for its development and cash potential. It was a complete turnaround in our traditional way of thinking. Yet so much of what we still do stems from the land; bricks, food, water and fuel all come from it. Houses, roads, businesses are all built on it. Viewing our landscape from the air, we see how town and country relate. Deprived border regions and prosperous urban hinterlands are all part of the same pattern. Fields appear as the cracked enamel surface of an ageing Chinese vase with thousands of hairline fractures. While our landscape has been shaped by human hand, it has also shaped us. Each subdivision tells its own family story. It's an unsystematic, irregular jumbled mess, probably not unlike how we secretly like to think of ourselves, in spite of all our recently acquired surface gloss and sheen. You'll never tame the Irish.

The vast majority of us no longer have any relationship with the land. We are a suburban society. Neither city centre nor country, much of the place where we live and work looks like the 'burbs: an agglomeration of housing and industrial estates, commuting routes and shopping centres. Urban and rural have now grown so far apart that there are in effect three jurisdictions on the island of Ireland: Northern Ireland; the cosmopolitan urban principalities of Dublin, Cork, Limerick and Galway; and then there's the rest. But 'the rest' defines in large part what Ireland is. What happens in 'the rest', the way the countryside develops over the coming decades, will decide everything from the quality of the water we drink to the health of our children to how we see ourselves as a nation.

This book is not one of those arguments that proposes rural Ireland as the true crucible of what it is to be Irish, that townies are somehow less innately Irish than culchies. Perhaps, though, it is time to ask how becoming an urbanised society has changed us and what we are turning our backs on. Have we lost our inner culchie? How did this happen so quickly? How has shifting, in two generations, from living with food rationing to 24-hour supermarkets changed us?

Where has our lust for land gone? What unseen changes has the Irish psyche undergone in the fifty years since Bull McCabe thundered at the American about the 'law of the land'? The only land that is hotly contested any longer is land banks on the outskirts of cities or sites with full planning permission, and even now that cash cow has been pretty much exhausted. Would Bull McCabe resist the American's plans to concrete over his field in modern Ireland? He would in his eye. He'd organise a rival consortium to drive the price up and retire to Marbella on the proceeds.

The countryside used to be the place of food production. An enormous part of every day revolved around getting it on the table. Our migration towards the towns has moved in step with a mind shift around food. Can you think of any other activity that you do at least twenty-one times a week? We commute, play sport, log on, phone our partners, shop, go drinking and, yes, make love a lot less frequently than we eat. Yet all these other activities are accorded greater importance. Food is something of an irritation, a necessary pit stop before we get back out on the track again. A chicken and stuffing sandwich munched between conference calls. You don't find this in other European countries, where a shared evening meal is a focal point of every day. Why are we so different? Is communal enjoyment of food something we have left in our rural past? In post-agricultural Ireland, has food been reduced to a troublesome bodily function? True, we have a new cast of superheroes: celebrity chefs and their apostles; the foodies. Yet there is a mismatch between what we actually eat and the gastro-porn served up on television. The evidence from the aisles of the supermarkets is that food made on TV stays on TV.

In increasing numbers we profess concern for the source of our food, but we have never been so disconnected from where and how it is grown. Modern farming practices have evolved over the last fifty years. That makes it a pretty young industry by any standards. Yes, intensive agriculture has given us unprecedented levels of production and cheap food. But it also gave us pesticide residues, herbicide poisoning, rancid lakes and rivers, antibiotic resistance and mad cow disease in humans. Ten thousand years of plant and animal husbandry has been rejected by the scientific and political establishment as backward. The only alternative to modern forms of farming, we are told, would be to take us back to the Dark Ages. Prices would go through the roof, shortages would lead to Soviet-era queues in the supermarket and our children would be struck down with everything from scurvy to rickets. Maybe they are right: but do we not already suffer from increased incidence of obesity, asthma and diabetes, at least partly as a consequence of modern diets?

The Irish countryside is up for sale. Who we want to own it, control it, plan for it and legislate over it is of huge importance, yet somehow currently beyond our range of thinking. The decline of farming, the value of land becoming its development potential and the urge to conserve the countryside as an amenity area for city dwellers are all changes that came about rapidly in Ireland, but are also things we let slip under the radar during the past ten years. We used the Irish landscape as a giant larder for making money—property developments, motorways and retail parks—while often the communities and traditional occupations of rural Ireland, particularly farming, were allowed to slide and decline in importance. And while we ate well, our knowledge of food and the basics of harnessing the land to sustain life have also taken a battering. In our race to be something other than what we were, Ireland lost its connection to what it is good at.

As food producers and small farmers struggle, the farms that survive into the future may soley be the ones that become bigger or are part of large companies that operate landholdings all over the world. Globalisation is coming to Gurranabraher just like everywhere else, and the question is can we, or do we want to, compete with the cheap food being made, often badly, somewhere far across the globe. The direction that rural Ireland will take is partly the responsibility of Irish consumers and the choices they are willing to make about supporting Irish food. The growing of food has been a good servant to Ireland. It has been our most successful long-term industry, one we are good at and one that currently offers real employment, penetrates vast export markets and has even bigger future potential. But if we turn our backs on rural Ireland and allow farming and locally produced food to decline, then big business and multinational supermarkets will dictate what we eat, and the traditional use of land in Ireland will begin to fade. The battle for who will control food, farming and the future of the countryside is currently taking place. Who wins that battle gets to write the future of the entire country.

Philip Boucher-Hayes & Suzanne Campbell
June 2009

PART ONE

Chapter 1 ⌒

WHAT'S THAT TERRIBLE SMELL?

Paddy Luxury

You know you're coming into Ireland when you descend into Dublin Airport on a rain-soaked plane and the man in the seat in front is reading the death notices in the *Irish Independent*. Something in your heart falls, while at the same time you get that warm, tea cosy-type feeling. The same feeling you get on hearing Gay Byrne's distant archived voice from his morning radio show, or mart prices for store bullocks being read out. It's the feeling of *old* Ireland. The Ireland before we all became rich and glamorous. Or, more accurately, before we wanted to be rich and glamorous.

The deaths page, the news of how we leave the world, not enter it, is of huge importance in old Ireland. It's very likely that the man in the seat in front knows someone mentioned in the deaths page. A neighbour, a teacher from his former school, the woman his uncle did a line with before the uncle became a priest and then came out as gay. Because somehow in Ireland we still all know each other, and as much as we may see ourselves as a fancy, multicultural race full of Eastern European girls with catwalk model legs, we are at heart a country where knowing someone, in some town somewhere, meant we were closer together. For better or worse: this closeness also brought with it valleys of squinting windows, streets of twitching curtains, informers, traitors, abusers and those who remained silent. But generally, it brought with it a shared cultural closeness, a feeling we were all in the same mess together.

But are we in the same mess any more, or have we split into tribes—urban and rural, Louboutin-wearing or Limousin-breeding? What is that divide and what has brought it about? Is it geography? Income? Are we a culture now separated along the lines of who did well during the boom and who didn't?

Where are the boundaries and crossing lines of new Ireland?

Until the 1980s, urban and rural populations in Ireland shared a common factor which has now almost disappeared: even in our capital city, most people were only one generation away from a rural family. This shared cultural currency was key to everyone's identity. Much as we might have liked to pretend otherwise, we had dirt behind our ears and welts on our hands. We had a connection to Ireland's countryside, we had knowledge of where food came from, of the cycle of life and death and the realities of rural life. The difference now is that we have a huge suburban population, born and bred in its own particular widescreen world, while in the countryside there are fewer farmers, more holiday homes, more urban and international migrants, and tolled motorways criss-cross our fields. We are a twenty-first-century country that has made the change from an essentially rural base in a very short space of time. Ireland has remade itself. We are mall shoppers, bling worshippers, M&S packaged food eaters and somehow strangers to our recent rural past. Have we have lost our inner culchie?

Perhaps the economic slow-down is making some of us reassess and reclaim our past. High food and oil prices and the rising cost of living may be bringing about a new austerity. The thing is, for Ireland austerity isn't new at all, rather a return to how we were living thirty years ago. The price of our over-commodified lives has made us reflect on how we live, particularly to examine what goes into our mouths and the environment that produces it. While Jamie Oliver and Richard Corrigan urge us to grow vegetables at home and Clodagh McKenna tells us to buy food from farmers' markets, the irony is that this is pretty much what everyone did until forty years ago, when all of a sudden we were told to fill that supermarket trolley, ditch the garden and stuff ourselves with food that had more air miles than Richard Branson and all the inner substance of Paris Hilton.

Supermarkets were the gateway to bad food. Convenience eventually led to cost-cutting, which eventually led to cheap-as-chips food, which in a lot of cases wasn't very good for us. As recesssion bites, we want food that costs less and less. And while much of what's in any supermarket is not going to kill us, handing over the control and, in many cases, the production and ownership of what we eat to enormous, vertically integrated corporations resulted in a huge remoteness from food, farming and the countryside that produces it. This four-decade journey of convenient, low-cost food is what brought us to *Supersize Me*, Gillian McKeith yakking on about stools and an obsession with food, weight and obesity. While eating that chicken pesto wrap for lunch might mean twenty rather than forty minutes on the cross trainer in the evening, constantly gauging what we eat against our body fat, exercise requirements and dress size is now a curiously normal activity in many people's lives. Not long ago in Ireland, getting *enough* food into your body was the challenge, especially when the average manual labourer burned about 3000 calories a day. What we deem to be a normal relationship with food simply isn't; it's a hallmark of the way we live at present, particularly since in most of the world, people are still actually physically active, and very frequently underfed.

EAT YOUR DINNER

Food security—simply having enough to eat—used to be something we all thought about. That's why our mammy told us to finish everything on our plate. She in turn had been told this by her mammy, whose own mother may well have remembered a time in Ireland when people actually died because they didn't have enough to eat. In terms of history, the death of over a million people from famine happened two minutes ago. It's a shocking reality that's hard to take in for most of us, let alone the next generation of orange-coloured Irish teens. In today's Ireland we never think about food security. We take for granted the 24-hour supermarket shelves and endless supply of food from 'out there' that we don't really think about. But what would happen if by some strange crisis the country became blockaded? What would we eat? How many of us know how to grow even the simplest of foods?

Take the example of wartime Britain, when the possibility of an island population starving was not far off the cards. With German u-boats blockading shipping, even Hyde Park was dug up to become a food-producing unit. Women and children became farm labourers and food was brought back into focus as the very stuff of survival. But in the post-war period, we allowed ourselves to become distant again from the very basics of living. Feeding ourselves is something that none of us is actually capable of doing, but it's an issue gaining new attention across the normally complacent developed world. Even Irish politicians are getting in on the act, having realised that the price of oil, the cost of transport and having enough to eat are three related concepts. In 2008, Trevor Sargent, the Minister for Food and Horticulture, announced a review of food security in Ireland.

Ireland now faces a similar challenge as Britain experienced during the Second World War, when its government realised there was a clear need for food security. They went from being 120 days self-reliant for food to 160 days. Their land under cultivation for food went from seven million acres to almost 20 million acres.[1]

Sargent's response to the issue was triggered after Feasta, a Dublin-based environmental economics organisation, urged the government to draw up a food security roadmap and an emergency food plan.

Food is (shock, horror, amazement) a limited resource, Feasta pointed out. It does not actually grow on the shelves of your supermarket but has to emerge from the ground somewhere and be harvested, transported and handled by real live people, until it arrives at where you purchase it.

> We've urbanised half the world's population without taking food into adequate consideration … we treat food as if it's another part of 'goods and services'. There's probably not a week's supply of food here for everyone in Dublin, let alone the country. The shelves would empty of food very quickly in the event of an emergency.[2]

This was the opinion of Feasta's Bruce Darrell.

So finishing everything on your plate, like your mammy told you, may not be as mad as it sounds. Shortages of oil, more expensive means of transport and less agricultural land under cultivation have had some Irish people thinking about actual food again: how it's farmed, how much it costs and where it's coming from now and in the future. And they're not alone. Many writers across the globe view this century as an era that will be characterised by resource wars as increasing populations compete for depleting supplies of energy and, more worryingly, of water and food. Perhaps this new focus will ask questions of how we relate to the Irish countryside and what goes on there.

If more of us are thinking about where food is coming from, we're also looking at what food is doing to us. In Britain, the sale of vegetable seeds surpassed the sale of flower seeds in 2008.[3] People are interested in the idea of growing their own food; perhaps they want more control over what goes into their bodies and are attracted by the prospect of paying less for it. On the other side of the coin are the large-scale commercial realities of producing enough of the stuff to feed the world at large.

As we become more food aware and more interested in slow food, it is the intensive agricultural practices that dominate farming, even in Ireland, that are actually supplying most of what goes into our mouths. At the end of the day, our demand for frozen pizzas is still beating our urge for organic rare-breed cuts of meat. While on the one hand we are told to live an idyllic grow-your-own lifestyle full of health, fitness and beauty, the reality is radically different. Food is now a game of two halves: on the one hand is the mass-produced commercial stuff that goes into your supermarket loaf of bread; on the other the posh minority of organically or artisan-produced food that is lifestyle-driven, highly priced and promises to change your life. It's like *X Factor* versus *Arts Lives* on RTÉ One. While we know we should be watching that documentary about John Banville, what we really want to do is zone out and look at Simon Cowell's hair, with a slice of pizza on the side.

However clean-living our aspirations might be in Ireland, less than half a per cent of what we spend on food is spent on organic produce. It's still commercial large-scale farming that is feeding us all. While many people won't agree that this is a sustainable or acceptable way to produce food, it happens to be the reality, but the future of the Irish countryside as food producer can include all guests at the party: green, organic, high-end and commercial farming. During the boom times in Ireland, when we thought building apartments was actual growth, the fact that we produced good food went out the window with the deaths page. But in times where Ireland's wealth and economic stability have taken a battering, a bit of refocus on farming and the food sector is not just a return to our past, but an investment in our future.

If we're interested in food, environmental issues and sustainability, we also might have to start re-examining the Irish countryside as somewhere not just of alien beauty to drive through, but as a place we can look at with fresh eyes and start a new relationship with. Or at the very least, we might need to understand a little more about food, how and where it's produced, and the people who live where it comes from. Rural ireland: the farmers, the Foxrock fannies, the Brazilians, the greenies and the priests. It's a picture of *us*, whether we like it or not.

WHAT IS THAT SMELL?

Meet John and Geraldine, two of the increasing number of urban dwellers with aspirations to the manor born. Not unlike many other successful city-based businessmen, John fancied himself as something of the country squire. Geraldine quite liked the idea of running a riding school. In the early 1990s they decided to get some real mud on the wheels of their 4×4 and made their dream move from Blanchardstown in North Dublin to a farm in County Kildare.

£1.7 million bought them a hundred acres, a stately pile, twenty stables and an indoor arena. Surveying his domain, John must have felt that his migration to the rural *haut monde* was complete. He had post and rail paddocks, reasonable-quality bloodstock and a bevy of underlings running around the place doing the work. Unfortunately for John, in 1997 he suffered the social and practical inconvenience of being chief suspect for the murder of Veronica Guerin. After a spell spent in another period granite pile, Mountjoy Prison, his present abode is Arbour Hill, where he is now serving a twenty-year sentence for drug dealing.

Doubtless John Gilligan made a pretty ridiculous spectacle strutting around Jessbrook Equestrian Centre during his happy period playing lord of the manor. Joel Schumacher's film *Veronica Guerin* contains a scene where John and Geraldine Gilligan are showing visitors around their paddocks dressed in what can only be described as equestrian costumes straight out of *Darby O'Gill and the Little People*: loud waistcoats, cravats and breeches abound. Perhaps the director was trying to make precisely that point; that John and Geraldine Gilligan were the *little people*, playing at being something other than they were, and looking fairly ridiculous in the process. But poking fun at violent criminals aside, it's worth noting that the pretensions of the Gilligans aren't much different from those of the large numbers of professional people who, during the boom years, pioneered their way from the 'burbs out to the Pony Belt; a land of electronically gated residences, private schools and a pony in the paddock.

North Wicklow, Kildare and Meath are full of urbanites who have the means and the motivation to move to the Irish countryside, and they have done so in droves. Alongside Eastern European farm labourers, they are probably the only section of our population who in recent years have bucked the trend of the urban migration figures. All of it, of course, driven by a very Irish dream of moving into the Big House and aping the mannerisms of the landed elite: 'Horse Catholics', as our grandparents would have called them. They inhabit rural Ireland's Wisteria Lane, where soccer moms feed from the trough of packed tables of goods in ever-expanding branches of Avoca Handweavers, picking at hearty country fare they don't have the time to make themselves because they're too busy lunching with their friends.

The attractions of country life are pretty obvious: escape from the noise and visual disturbance of traffic, stressed-out supermarkets and crowded streets; the beauty of green, unspoiled rural areas and small, cutesy villages. So the myth goes. However, most of those who can afford to make the move to a country life within commuting distance of a major city are not living in an undesirable city centre location in the first place. Far from the scenario of swapping a grimy apartment on Gardiner Street for a cottage near Enniskerry, those who can afford to make this move are more likely to have sold a five-bedroomed house in an area such as Foxrock for that cottage in Enniskerry.

Needless to say, they'll be adding a four hundred square foot kitchen, two more bedrooms, three more bathrooms and a hot tub on the deck. But they'll keep the paddock. After all, Imogen is doing very well at her riding, it's improving her confidence at school and it's time she had her own pony to share horse-riding play-dates with high-status child neighbour Daisy.

But this is where the rural idyll ends, because although he may drive a Land Rover at weekends and keep a black Labrador, our Mr Pony Belt, the immigrant father from Foxrock, still commutes daily to the offices of his frazzled, depleting workforce in the IFSC. As for his wife, life hasn't changed too greatly either, except for the exasperating half a day she now spends behind the wheel of the BMW X5 because there aren't any schools locally where they feel entirely happy about sending Imogen and Jacob. ('Ohmygod, that *dreadful* local accent!')

Mr and Mrs Pony Belt would get involved in the community, but they just don't have the time. And when they do … well, let's face it, you can only really pick up every third word of what the locals are saying. For this family, being within twenty miles of Dublin is an essential part of the rural life package. Any further away and Mrs Pony Belt would suffer separation anxiety from BT's designer floor, Donnybrook Fair and her South Dublin Pilates class. Her hair colourist, waxing and nail technicians almost fainted at the idea of her even thinking of going to a salon outside Dublin. For Mrs PB, anything beyond twenty miles from the city centre is bandit country, far beyond her health and beauty comfort zone. Yes, okay, times may be tougher now, but 'like, ohmygod, where would you have your Shu Uemura treatments'.

Unless you've a helicopter. The PBS have thought about it. The licence only takes forty-five hours of training, but Mr PB's diary was too busy to continue his classes at Weston Airport. He thought about sending his PA to do the exams in his place, but then decided on a new Aston Martin DB9 instead. That was before he realised the true extent of traffic and obstructions on rural roads. Like tractors! Why do they go so slowly? Why aren't they banned from the roads in the early morning hours when real people like him have to get to work? Not only that, but there's some crazy person on Mr PB's route who stops traffic to bring a herd of cows across the laneway at seven a.m., just when he's turned out of the bottom of the drive. Every morning, Mr PB contains a Jeremy Clarkson-like explosion as he pulls his Aston to a halt in front of the cows, telling himself to breathe deeply and concentrate on the smooth walnut interior of his 5.9-litre beast. The cows (why are they always black and white, aren't there other colours?) gaze gloomily at the car as very slowly, udders swinging, they amble towards the other side of the road.

The Aston roars on towards the N11 and town. Cursing his wife's rural aspirations and the cost of filling the tank, Mr PB is beginning to realise the inconvenience and possible madness of living in the country. Like the incredible smell from the surrounding fields at certain times of year. Someone in his office said it was shit from animals spread as fertilizer. Ban it, for God's sake, the stuff stinks. And the shops, don't get him started ... the local newsagent doesn't even stock *Men's Health*, and trying to find a decent skinny latte is like searching for weapons of mass destruction. On top of that, there's nowhere to run or go to the gym. The fields have uneven terrain and allegedly the farmers will shoot you on sight if you run over their crops. Food is another huge problem: his wife loves ambling around farmers' markets, but neither she nor the kids will eat vegetables that look as if they ever had dirt on them. Everything that goes into their mouths comes out of plastic packaging, and is calorie and allergy labelled. Let alone sourcing the kids' vitamin and health store supplements and the white tea (yes, white) that Mrs PB has become obsessed with. No wonder they spend most of their time going in and out of Dublin.

Mr Pony Belt is becoming nostalgic for Friday nights in the Harbour Master; pints and after-work banter with the guys; the good times when hedge funds rose endlessly and his wife hadn't heard those two f**king words *River* and *Cottage* in the same sentence. How on earth did he end up living in some field in the middle of the country? Even the pony they bought for Imogen has turned into an unmanageable terror. Knowing nothing about horses and trying to placate a small but very persistent child, Mr PB tried his best by phoning a local riding school and asking where he could buy a Tennessee Walking Horse (Imogen's favourite from her *Incredible Book of Horse Breeds of the World*). The riding school people (he had a sense they were holding in laughter) told him that Tennessee Walking Horses are more common in Tennessee than Wicklow and told him to get a small manageable pony instead.

The small manageable pony arrived; he has eaten them out of house and home. He needs new shoes every six weeks, vet visits a couple of times a year, a €1,200 saddle and a €4,000 horsebox. Imogen rode him once and he bucked her off. What followed was a horrific journey with a screaming child in full traffic to Beacon Private A&E. (No way were they going to their nearest hospital; it used to be a workhouse, for God's sake.) The pony was sold to the riding school. Imogen cried for a day, but good counselling always sorts these things out.

PEACOCKS ON THE LAWN

Mrs Pony Belt is now looking for property in France. Saint-Paul de Vence, she thinks. Good sunshine, no bad smells and decent shopping; Mr PB could fly up and down from Nice; leasing a private jet would be a smart idea. She's tired of people telling her there's a recesssion on. I mean, once you start out with a high-end lifestyle you can't really give it up, can you? People think you might be feeling the pinch. Unfortunately their cottage in Enniskerry will sell for less than they paid for it, but on the up side, good property in France is also tanking in price. After all, she's heard the rumours that the Doyles can no longer afford their place in Villefranche; perhaps she might put in an offer. It's a bit cheeky but who gives a crap? Getting hold of good property is dog eat dog, darling. Though she will miss the Gordon Ramsay at the Ritz-Carlton at the top of the village, and that lovely equestrian shop where she gets her Barbour scarves and Harry Hall jodhpurs. But when you've got a Fragonard in Saint-Paul, how can one compare?

So our Pony Belt family found rural life not all it's cracked up to be. For those seeking sushi and Hunter wellies all in the one package, living in the countryside will most likely fail on both counts. It's dirty, has unusual smells and the inconvenience of being at least an hour's drive from a Michelin-starred kitchen.

But some other new countryside migrants show more commitment. They seek a rural living space further away from big cities, and full-time immersion in rural life. In counties such as Carlow, Kilkenny and Cork, remodelled farmhouses with underfloor heating and Agas have become the sought-after homes of those fans of *Country Life* who also occasionally read the *Irish Farmers' Journal*. This demographic know the difference between hunter trials and a three-day event. They consider themselves a little more knowledgeable and also demand a little more of their rural experience. Not content with owning the most fashionable dog (a rescue), this group feels they are rural-aware and green-aware. Rural Ireland is where they want to live; quietly, ethically and as part of a community of those similar in outlook. But again, their search for a country life experience is a particular version of rural Ireland, which is of course a very narrow slice of the story.

Like Mr and Mrs PB, they have made an escape from suburban life to a dream rural farmhouse, with gardens, a heron pond and a jasmine-clad porch at the heart of a dozen or so post-and-railed acres. Mature trees and a river stocked with trout complement the property, but the only livestock they'll keep are a few rare breed Jacob sheep to keep the grass down. These Country Lifers frown on commercial farming. They may buy their meat in the nearest farmers' market, but they prefer not to think exactly how it was killed. Their rare breed handbook and web groups gave them all the advice they needed to start up their flock. They say it's just a hobby, really: Tamworth pigs were also in the plans before they found out they make your pasture look like the Somme. They've stocked their limestone out-sheds with Silkie hen pullets, bought mallards to live on the pond and a pair of peacocks (for that Elizabethan look), before their screeching calls had them leaping five feet out of the bed in the middle of the night. There are composters in the garden, wind generators planned for the roof, and green issues and conservation are much talked about around the scrubbed pine breakfast table.

Hobby farms became the top status symbol for both green-minded urbanites looking for escape and some of Ireland's millionaires, who in the past fifteen years sent the price of agricultural land spiralling. In 2007, the value of Irish farmland was heading for €60,000 per hectare (€24,281 per acre), the highest in Europe. At the time it was almost ten times the value of land in Scotland and six times that of similar farmland in England. Many estate agents pointed out that hobby farmers and wealthy urbanites were pushing up land prices far beyond what regular, food-producing farmers could possibly pay for it, with predominantly 'high net worth' individuals accounting for sixty per cent of farm purchases in 2006. 'Serious farmers who want to engage in farming on a commercial basis and earn a decent living are now looking at buying farms abroad,' said Derek Brawn, Savills's then Head of Research.[4]

Whether it's Michael O'Leary's herd of Aberdeen Angus or celebrities selling us the attractions of growing your own vegetables, the idea of pretend-farming is a desperately attractive one—all the glamour of prancing about one's land without any of the stress of having to make any money from it. And it's no surprise that from the late 1990s the phrase 'hobby farm' was eagerly set upon by bevies of excited estate agents eager to attract buyers to the reams of ill-defined lands and tumbledown houses sitting on their books. During the crazy years of high property prices, the term was tacked onto any sort of land under fifty acres that agents think might attract urban dwellers with fantasies of *The Good Life*. Plots are typically one to five acres, the lands with attached properties are most desirable, and being within striking distance of a large town or city ups the price bracket considerably.

Hobby farming is a trend that's grown enormously in the UK in the past decade. In 2006, the Royal Institute of Chartered Surveyors pointed out that nearly half the farms sold in Britain were bought by non-farmers. They described urban dwellers who'd shifted their lifestyles to houses in the country with twenty to sixty acres of land; larger than a garden but not a fully fledged farm. This trend sits alongside the huge decrease in the number of people working as full-time farmers in Ireland and the UK. In Britain, smallholdings now account for approximately six per cent of the agricultural area of England, and in parts of the south and west over ten per cent of the land area is now used for smallholdings and hobby farms.[5]

Part-time farmers still account for most of the farming activity in Ireland, but the hobby farmer is a new arrival on the scene. They may come from a rural background, but hobby farmers are more usually urban dwellers and weren't reared on a farm. Farming for them is a pastime rather than an occupation. And as small mixed farms have all but disappeared from Ireland's commercial farming scene, hobby farmers are taking up residence in precisely those smallholdings that are useless to those working in the reality of modern farming. Most have the type of acreage that can't be exploited for large-scale cereal growing or milk production (where hundreds if not thousands of acres are now necessary to be competitive) and many have marginal land, or land with trees, ponds, etc., which are only attractive to people who have the luxury of not having to make any money from where they live.

The phrase 'hobby farmer' has also been used to describe those who farm small quantities of produce for sale in farmers' markets. Many feel this is a misnomer; 'hobby farming' suggests that your expectation is to make no money at all. Author Willy Newlands, who penned one of the recent crop of how-to books,[6] suggests that anyone moving to the countryside is mad if they think they can make a living out of less than fifty acres. He points out that inputs, stock prices, etc. take up virtually all of any margins there might be. Organics offer the obvious attraction of a higher margin for any produce that hobby farmers are planning to sell, but taking account of the fact that they are more labour-intensive to produce, and that the market in Ireland is full of cheaper imported produce, it's a game not to be depended on for those in small-scale production.

The attractions of keeping a couple of Gloucestershire Old Spot pigs around the place and selling the pork is one of the most frequently professed hobby farm dreams. It's agreed by all that rare breed pork is pretty good to eat, but the reality of three sows and their progeny eating as much as twelve tonnes of commercial food a year puts another, more scary, spin on the story. Livestock also needs a fair bit of hands-on expertise, daily care and, at the end of your beautiful relationship, a trip to the slaughterhouse.

This brings us back to the problem that many urban people have with farming of any type: the unpleasantness of much of its basic realities. This has led to tension in some rural communities and, at worse, complete breakdown between newcomers and those who have been there for generations, and who are active in farming or hare coursing or doing whatever it is that drives the newcomers mad. Some say that far from enlivening the countryside with new blood, hobby farmers and Pony Belt immigrants have created many divisive problems. The most vociferous objections to hunting, large-scale commercial farming and the spraying of fertilizers and pesticides frequently come from this group. Farmers complain that newcomers don't understand the real business of making money out of farming; it's easy to turn your nose up at a slurry spreader in the fields next door to your house if you don't have waste from bovines to dispose of or top-quality grass to grow. These activities are what makes up farming itself and provide farmers with their income; farmers point out that the newcomers' incomes come from elsewhere. The two populations sharing the one area essentially become two peoples divided by a common language.

WHAT AN *APPALLING* BUNGALOW ...

The biggest point of tension between the new and old rural dwellers revolves around the issue of land in the countryside: its price and, more importantly, control of who lives there. Whether they live in the Pony Belt or deep in the countryside, or own a holiday home, the migratory middle classes seeking bucolic refuge pushed the price of land beyond the reach of any but themselves. For example, most rural dwellings with more than half an acre in North Wicklow were selling above the €1 million mark until prices finally reached a tipping point in 2007.

Even in a falling market, most of these dwellings are far out of reach of those who work in farming or the rural economy. Not only that, but opposition to one-off rural housing from urban people now living in the countryside is an issue leading to particular ire. In their view, once the dream dwelling is found, why should the pretty country road approaching it be ruined by the blight of *horrendous* bungalows or, more recently, Palladian redbrick giants (definitely not the type featured in *Country Life*)? They cite cases of houses not built by or lived in by people who work in the rural community, houses that are merely cash cows to be erected, sold and money banked by greedy landowners or developers with a golden telephone hotline to the local planning office.

This gap between traditional and new arrivals in rural areas reaches its widest point over planning issues in areas of natural beauty that carry a heavy premium for tourist and part-time holiday homes. Donegal, West Cork and any area where holiday homes have mushroomed in vast numbers have been loci of discontent from locals who feel that planning should be focused on local needs and not only on homes for those who only live in the community for a part of the year. As the value of property has fallen across the country since 2006, many rural dwellers feel relief at the prospect of more housing becoming available in their budget range as the rush to buy one-off homes as part-time residences or holiday homes is slowing down.

What's clear is that in modern Ireland there are many different types of people with different needs living in the countryside. It is no longer a monoculture of farming alone and a population made up of people involved in food and farming businesses, or in towns providing services to hinterlands occupied by farms. Changing trends such as the development of suburbs outside country towns, less one-off housing, more commuting and the urbanisation of rural areas have meant there are many competing views of what the Irish countryside should be. Is it land on which to grow food, or a pastoral landscape that should be frozen in time for urban dwellers and those lucky enough to live there on a part-time basis to enjoy for its beauty and uniqueness?

In a world where rising food prices and food shortages are part of our future, the emphasis on the countryside as somewhere to grow food turns the perspective back towards large-scale commercial farming and the benefits that it brings to the agribusiness sector. But in a new climate, terms such as 'sustainability', 'environmental protection' and 'carbon-reducing practices' are now part of farming and are not likely to go away. Irish farming can profit from huge demand for food products and rising prices around the world.

With the failure of the World Trade Organization talks in 2008, Irish farmers have more potential to control what happens to prices and, particularly, the future of beef farming in this country.

Rural Ireland will benefit from the huge global appetite for food, but alongside this is the pressure to produce the stuff in a more environmentally friendly way and knitting together rural communities which have altered greatly in the last twenty years. The question is, having moved from old Ireland into a very new era, where does the future lie for rural Ireland? Is its importance lost to us as we become more urbanised and lose our connections with how food is produced and with the basic realities of farming?

Rural Ireland is a place that means many things to many people. What this book explores is how we relate to it, treat it, eat from it and what its future might be. As the number of people in the countryside involved in full-time farming shrinks to 40,000, is rural Ireland itself becoming a suburbanised way of life increasingly removed from an old-fashioned sense of community? If how we use the land changes, do we change with it? Will the pressure to feed ourselves tip the weighing scales back in the farmer's favour? And if it does, will modern agriculture, with its industrial-scale farming practices, win the battle to produce food in the most cost-effective way possible? Can our health, our wallets and the environment afford for farming and mass food production not to change their ways?

Lines have been drawn in the struggle for which set of competing interests gets to run the country and a meaningful portion of our lives. The battle has begun and in a generation the winners will have reshaped Ireland and the Irish, how and where we live, what the country will look like and how healthy the people who inhabit it will be. This book examines the skirmishes taking place today and how we will be affected by them tomorrow.

Chapter 2 ∾

WHO DO YOU THINK YOU ARE?

Dude, where's my inner culchie?

It used to be that you could tell a culchie from a townie at twenty paces. Townies ate lunch; culchies sat down at noon for 'the dinner'. Townies walked to school to the De La Salle; culchies got the bus to the tech. Culchies ordered Guinness and a packet of cheese and onion Tayto; townies, Carlsberg and a packet of dry roasted. Culchie weekends were spent eating hang sangwidges out of the boot of the car at a GAA match; townies would suffer interminable drives in the country with their parents. Culchies were usually farmers. They hung around with sheep, knew what a Simmental was and occasionally smelled.

On the football pitch, some claim, a townie could be spotted by the way he flicked the ball off the ground with a practised jab of the foot. 'F**k that fancy shite,' said the culchies, 'go down on the f***ing thing.' Culchies had an uncanny ability of predicting the arrival of rain to within two minutes, but still hung around in it to get wet, whereas townies brought a coat 'just in case'.

We had scores of tell-tale signs that revealed, as clearly as if you wore brogues or hobnailed boots, which side of the town's speed limit signs you lived on. A mineral versus a Coke. Ice cream and jelly versus sponge and custard. In culchie Ireland, 'the lads' is a phrase that also refers to girls, the *Irish Farmers' Journal* is read rather than the *Irish Times*, and pretty much every person, house and piece of land within a ten-mile radius is known in intimate detail. In the North it used to be said that all anybody needed to know to pigeonhole you was where you went to school. In the South, throwing in a few jive steps on the dance floor would be enough to betray your background. These things mattered because we all looked down on culchies.

Dubs born within earshot of the bells at All Hallows would slag off anybody from far-flung Balbriggan as a 'bleedin' bogger', because even contemplating that anybody could come from further afield hurt their souls. Not only that, but we in Ireland created a peculiar pecking order to defining culchies. If you came from somewhere like Wexford town, people from Enniscorthy were culchies. To the townsfolk of Enniscorthy, the villagers of Camolin were the real culchies. And in Camolin, the people from five miles outside the town were culchies. And for that matter, using the word 'townie' at all is something that only culchies would ever do.

The popular school of thought in Ireland is that all these differences are now dead. We are the same homogeneous lumpen mass, slavishly following imported fads. Whether your teenage daughter is from Ballymun or Bagnalstown, she will speak like a Valley Girl, or try to. All our aspirations are the same, no matter where we are from. But is this really the case, or does this school of thought look at Ireland through a very narrow prism? Might there still be a world of real difference out there? Perhaps culchie Ireland is still alive and well but simply hidden, ignored by an increasingly globalised culture and a media that is largely based in the east of the country. It's possible that over the past ten years we forgot that culchie Ireland is still out there. Maybe now it's time to examine it again, take it out and give it a polish, and look at the reflection of ourselves and our past. While we tried to remake ourselves in Ireland, buying houses in Bulgaria and Jimmy Choo shoes, are we still a nation of culchies at heart?

WHAT DOES IT ALL MEAN?

Not long ago in Ireland, to describe someone as a culchie was to 'out' an individual as a knuckle-dragging, one-toothed hick, newly escaped from the Planet of the Apes. As in, 'Looka da bleedin' tick; he's only a bleedin' culchie.' Those on the receiving end of the term tended to respond in one of two predictable ways:

1 Being called a culchie was something to be proud of, as it meant you were not, in fact, a pig-shit-smelling yokel but a 'cultured person'. Or,

2 That the person who had called them a culchie was nothing but a witless skanger from whom you'd expect no better, being of such low birth and beneath the dignity of response.

A bit like the rebirth of the word 'queer', Ireland has recently come to embrace the term 'culchie'. We have culchie festivals, awarding the title of Culchie of the

Year to the best mucky-faced, flat-cap-wearing male. The origin of the word is much disputed. Some say that it derives from the town of Kiltimagh in Mayo. It's certainly Hiberno-English, the particular brand of English that we peddle in Ireland, which has exported words such as 'craic' and 'shite' to such RP media as the BBC. It's interesting that the word 'culchie' has probably been used less in recent years—Ireland now has many new identity and multicultural crises to attend to. Why slag off culchies when there were shedloads of blacks, chinks and Poles to have a go at? Even the poor gays have slipped down the unpopularity ladder: 'I mean, at least the queers are bleedin' Irish, wha'?'

In present-day Ireland we are clearly wrestling with the new version of where we live. In the past thirty years we have dragged ourselves from a monochrome Ireland to one more resembling the United Colors of Benetton; multi-accented, multi-ethnic and mixed-race Irish people now populate both our streets and our fields. The importance of being either a townie or a culchie was slipping. While generations of Irish tried desperately to escape the place, for the last decade or so it seemed the whole world wanted to live here. Work was plentiful, but once the economy took to its sickbed and many immigrants have returned home to Poland, Latvia and Brazil. Others have stayed, enjoying the Irish way of life and marrying into Irish families, God love them, and yet others will continue to arrive. Geneticists might say that the new racial mix will give Ireland some hybrid vigour; a mixing of bloodlines that could improve the general appearance or talents of our national stock. We too could produce a Helena Christensen, an Alicia Keys or a Tiger Woods. Looking at archive film of toothless, wide-eyed Paddies of the past, some say the racial mix can only be an improvement. At worst we'll end up looking like Labradoodles.

NELL MCCAFFERTY'S CIGARETTE

As we're now allegedly a grown-up nation, a nation that partied through the economic boom and then hit the inevitable hangover, are we better aware of our identity, and happier with ourselves, and others, as being Irish? Thirty years ago, Ireland, especially when viewed from afar, was a green and rural land. Men wore flat caps, played accordions and stood about on street corners after Mass smoking and talking about the 'awfa' poor stay atha' neighbour's cattle'. We had old-fashioned values, a dominance by the Church and a strong culture of musical and literary traditions. This was a place where communities were tight and rural experience was strong. Music folklorist Davy Hammond described 'leaning over the half-door, dozing and dreaming'. A collector of folk traditions and songs, he was a man brought up in 1930s Ireland, where half-doors functioned to keep four-legged occupants out of the house and not as photo-ops for tourists. This was an Ireland where a neighbour would spend several hours at your house chatting, before finally working up to ask for the lend of your spade. Talk was important and there was plenty of time for it; culchie Ireland was alive and well.

This way of life could still be seen in images we produced of ourselves even up until the 1980s; John Hinde's tinted postcards, Radharc films about basket-weavers and Nell McCafferty waving a cigarette about on the *Late Late Show* complaining about the Church. Even Dublin, with its cattle marts on the streets and dairy farms on the fringes, was a city with strong rural ties. Many of Dublin's people were themselves country people. Inheritance structures meant there were plenty of Irish sons and daughters without land, businesses or trades to inherit, many of whom made their way to Dublin in the forties and fifties, working as secretaries, clerks or labourers, or training in the trades. But the culchie gene remained strong. Returning to rural areas in the summer, they still brought in the hay with everyone else, drinking tea mixed with milk in glass bottles that the women brought to the fields in the afternoon. News from home and your county's progress in the Championship were often more important than what the city had to offer. But the city had jobs and opportunities when rural Ireland did not. The flats of Rathmines, Ranelagh, Drumcondra and Phibsborough were full of people who were essentially from somewhere else. Perhaps we were then, whether town or country folk, all essentially culchies; frayed about the collar and with plenty of dirt under our nails.

If these images of Ireland, particularly as a rural place, are clearly of our past, what kind of people are we now? This is an increasingly tricky question. In the new physical and ideological map of Ireland, the one clear trend that can easily be seen is the rush by both rural and urban Irish towards a certain kind of homogeneity or sameness which can be best expressed in one word: wealth. If our economic boom encouraged all of us to reach for our slice of the pie, it also changed the way our communities and social structures work. Getting rich became important and, possibly, accessible. How quickly we moved from dirt to Donna Karan.

In the boom years, Ireland changed from a place where class, community and sense of place were very clearly marked out to one where these differences became less important. In the past there were layers to how our communities functioned: people occupied specific places, defined by their background, level of education or skill, and religion. You were either a teacher, priest or shop owner (high status) or a delivery man, farm labourer or street sweeper (low status). And that was pretty much your lot for life. In today's towns and villages we could argue that the most important definition of anyone's place in a community is not their origin or profession, but simply whether they are rich or poor.

Some would say this is a good thing; there is no upper or lower class. Society now is a meritocracy: if you work hard and have a little bit of luck, you too could have a swimming pool in your back garden and a home cinema in the basement. In the new Ireland, those without the right education or background could become millionaires; meanwhile, the family from the Big House up the road (top of the pile circa 1929) have finally sold the place and are auctioning off the furniture to pay the bills. It's come full circle. We're all the same now, aren't we? Furthermore, if we get rid of the Church as the defining social arbiter, there are only the boundaries of capitalism and consumerism to contend with. Morality, community, family are sideline choices: you can pick and choose which code to follow, or not choose a code at all.

Many people feel that the biggest forces in our society today are to make money and to be successful. One of the most frequently asked questions in Ireland used to be 'Where are you from?' Even before the boom years, it had been pretty much replaced by 'What do you do?' In old Ireland, asking somebody where they were from was an essential part of filing their identity in the mental Rolodex we stored in our head, full of names, counties, pubs, townlands, football teams and a host of random countrywide acquaintances or family that could link you, however tenuously, to a person, wherever they came from.

'Ah, you're from Ballyshannon? Do you know any of the McGuires who run the undertakers there, just before the bridge?'

'Sure, isn't she married to Paddy Hearn who used to be the teacher up at the school?'

You know the drill.

Is what county you are from still important? Are place and community and tribe still of value in new Ireland, or have wealth and status become the new way of tagging people? Do we care about the ties and connections we have with people as a nation, or do we just care about ourselves? Did cleaning the dirt from under our nails make us any happier?

THE SACRED HEART

Let's remind ourselves of the Ireland of our parents and grandparents, when no one had heard of extra virgin olive oil or biofuels, and things were on the whole much simpler to understand. Josef Locke is crooning 'Dear Old Donegal' on the radiogram, its low, sonorous buzz lighting up the sepia tones of the family sitting room. There are flying ducks on the wall and maybe a portrait of the Sacred Heart, the flickering red bulb at its base showing the way of truth and light to all good Catholics. This was a time and a place where everyone in Ireland knew pretty much who they were and where they were going.

By the late 1960s, Donagh O'Malley's Free Education Act had introduced secondary education for all children. If you were a very good girl or boy who achieved four or more honours at Leaving Cert, a university scholarship might come your way, but for most Irish families, keeping a child at school that long, or sending them to university, was far beyond their means. Entry to the professions (law, medicine, etc.) was limited to those who could afford it. The possibility of earning beyond your family's potential was limited, and children frequently emigrated abroad to support siblings or parents at home.

Most people in Ireland were pretty sure of their identity as Irish, and as coming from a particular county or town. Beyond that there was religion, and class. Like tended to hang around with like, and everyone knew their place, from their *bench* or *pew* in the local church (words that in themselves denoted your religious background) to whom it was appropriate or inappropriate to socialise with. The class system in Ireland was an unwritten but complex set of rules, upheld by all its members, and those attempting to jump rank faced penalties of ostracism.

Todd Andrews writes of Dublin in 1907:

> ... at the top of the pile were the medical specialists, fashionable dentists, solicitors, wholesale tea and wine merchants, owners of large drapery stores and a very few owners or directors of large business firms ... at the bottom of the heap were the have nots of the city, consisting of labourers, dockers, coal heavers, messenger boys and domestic servants.[1]

In this slower, pre-cappuccino-maker Ireland, at the top of the pile in rural areas and urban townlands stood the Big House. Often remote from life in the surrounding village or town, the Big House was still a locus of curious glamour combined with, for some, annoyance at its silent granite symbolism of the Crown. In the post-war period, the houses that still doggedly survived were a source of discomfort to many Irish people. While estates provided employment and a marketplace for services, they were often seen as an Anglo-Irish entity, as something that hampered Ireland's development of its own agricultural sector—a belief that was handy for nationalist ends but was very often not the case.

BAD HOUSEKEEPING

The Big House, and its journey from seat of the landed gentry to its recent reincarnation as five-star hotel, mirrors our own transformation from a people who could cut turf and describe a blackthorn hedge to a nation whose aspiration was to mix good martinis. These days, many of the Big Houses are swathed in golf courses and owned by international hotel chains, their doors flung open to descendants of people who were might have cut their lawns but were clearly not allowed to walk up the steps and in the front door. 'Cúl an tí', meaning the back of the house, could be another clue to understanding the origin of the term 'culchie'. Indigenous Irish were seen as people who worked for the Big House, their place firmly out of view. Now we are a people who enter by the front door. The journey of the Big House from Anglo-Irish demesne to Ritz-Carlton hotel is one that strangely mirrors our own path to maturity, if we can call it that.

From their inception, country estates were simply lands tamed into submission by a predominantly Anglican elite who arrived in a terrifyingly strange and threatening Ireland only recently won and insecurely held. Many of these houses were built on top of earlier existing fortified dwellings, the newer constructions tending to be extravagantly built and modelled on classical European buildings. The Georgians, particularly those of the Regency period, were the kings of bling when it came to the country house. Extravagant gardens, follies and ornamental lakes were commonplace. It's hard to picture the physical graft involved in landscaping grounds such as those at Russborough House with nothing more than manpower; the scale and amount of labour necessary to achieve it is staggering. Digging ornamental lakes by hand is only possible if you happen to own a bevy of slaves or have a plentiful supply of grateful, but heavily worked, cheap labour.

The Big Houses were mini-economies and training centres for skilled workers, stonemasons, gardeners, horsemen and craftsmen such as coopers, blacksmiths and carriage makers, all of whom had their place in providing services for and gaining income from the operation of these estates. Unfortunately, many were badly, if not catastrophically, run. Nearer the time of the famine, it didn't help that some landlords' lifestyles far exceeded their incomes. One such was John Scott Vandeleur, who fled the country after gambling away his estate in Clare. Others incurred or inherited debts from bad management: by 1838, Lord Middleton had a nightmare £70,000 in arrears on his Cork estate. Examples of such shoddy housekeeping fostered the belief that landlords as a whole were an incompetent and cruel set of wasters, an essential element in the nationalist mindset. After all, it is easier to legitimise armed rebellion to get your land back if the people in control of it are making a complete pig's ear of it.

The famine of the 1840s was the trump card in this particular game. It was to become a defining feature of Irish identity, particularly the notion that rural policy and the mismanagement of the crisis by the landlord class and English administration were at the heart of the tragedy. Without the famine, we'd have a smaller stick to beat the Big House with—after the famine it could more easily be characterised as the devil's lair of bad practice, neglect and the source of much cruelty visited on rural populations. As with many dogmas, this didn't happen to be the case. The famine resulted from a combination of factors and, annoyingly to those still intent on blaming the Big House, focusing on the part it played is a waste of bile, for as Roy Foster pointed out, disappointingly for some, 'landlords are not central to Irish History'.[2] But many of us were not to know this then. We shook our pitchforks in the direction of the Big House; the culchie Irish were right and the robber baron English—driving us towards starvation, delighted to have us out of the place—were wrong.

By the late nineteenth and early twentieth century, the Big House party was drawing to a close. Land agitation, electoral reform, the Great War and the establishment of the Free State brought many estates to their knees. It didn't help matters if, like Bellevue House in Wexford, your house was burned to the ground and you were left without any option but to leave. Many Anglo-Irish families did pack up and move on, leaving behind them crumbling granite and briar-choked gardens. And who better to take over many of the Big Houses than a waiting army of people who, like the early Georgians, were fans of expensive furnishings, over-the-top outfits and 'my velvet shoe shan't touch the ground' behaviour: the Catholic Church.

Drawing rooms became bishops' parlours, housemaid accommodation became dormitories for errant young men. Many Big Houses, such as Moore Abbey in Kildare and Edgeworthstown House, metamorphosed into schools and religious institutions. The land acts and the Land Commission accelerated the break-up and redistribution of their surrounding lands, transferring over 13 million acres to small tenants, altering the shape of the landscape by creating hundreds of small farms with a central homestead.

SWAROVSKI-LIT SWIMMING POOLS

But in the last thirty years the houses themselves hit another rut with the shrinking of the Catholic Church and its religious institutions. The reformatories for boys are now closed, the palaces struggling on their last legs. Many Big Houses have come full circle and became rebirthed into the modern incarnation of the big hotel. For the first time, Irish people of even the lowest doorstep level could access a weekend away playing lord of the manor. You too could ride, shoot, fish and sit at a mahogany dining room table made during the Napoleonic Wars. Children who might have once peered through the murky windows of a boarded-up building, such as the Lough Rynn estate in County Leitrim, now had the chance to walk the hallways, admire the stucco and, if you had a few quid, buy a house on the estate. Failing that, you could always play a round or two beneath the parkland oaks, planted for very different reasons than sheltering golfers.

The glamour of the Big House is evidently still with us. Its allure still pulls us through the doors to enjoy over-rich meals, spa treatments or to own a piece of the dream ourselves. During the boom years, the culchies left the back entrance far behind and marched through the front doors in droves, handing over credit cards for afternoon teas costing €50 a head or shelling out what their dad used to earn in a month for a pint of stout. In June 2008, a full-page advertisement in the *Irish Times* Saturday supplement announced the opening of the Cedar Club at Powerscourt House, now a Ritz-Carlton-managed hotel. In what is basically an ad for a gym, the elaborately written copy implores a select few (250 people) to take out membership in the Cedar Club, which boasts 'ethereal light, and a Swarovski-lit pool'. There, you will find a 'lifestyle of luxury … while being enveloped by the majestic Sugar Loaf Mountain'. The text reminds us a few more times that the Cedar Club is 'truly for the privileged few'.

Three days after this advertisement appeared, on 24 June, the ESRI published their first 'R' word report, sending politicians, economic commentators and radio and television presenters into a paroxysm of despair and frenzied blaming over the approaching recession. While being enveloped by the Sugar Loaf Mountain sounds like an interesting James Bond plot, it's unfortunately not enough to save your job, thirty of which were lost at the aforementioned hotel in the weeks following the publication of the ad. In the following months, several Irish luxury hotels closed their doors—the glamour bubble had been firmly burst.

Many pointed out, like Chicken Licken, that they had told us so: once you get that rich and silly, it's only a matter of time before the sky comes falling down. But the journey from the Big House to the big hotel says a lot about ourselves, our habits and our changing attitudes to wealth and class. If landscapes and the spaces we live in 'are signifiers of the cultures who made them',3 what does this say about us? What does the Big House and its surrounding golf course say about the changes that have happened to culchie Ireland? If we could take a giant aerial photograph to reveal our cultural values and our way of life, it's a very different picture than one that might have been taken even as recently as thirty years ago, when small towns, regional cities and even Dublin were still surrounded by satellite farming villages such as Skerries, Rathfarnham and Rathcoole. The image then would have been of an Ireland where rural and urban culture were often closely intertwined. If the contemporary snapshot of Ireland is crisscrossed with motorways and huge suburban conurbations, where the new demesne is the shopping mall and its car parks, the old demesnes our luxury hotels and playgrounds, is it a good picture of ourselves? Perhaps more important, is it a picture of a society that is working well?

YUPPIE KVETCH

So we all got rich and spent weekends away in the Big House where granddad used to be a cooper's apprentice before he got the arse beaten out of his trousers for stealing an egg. Poor granddad, if he could see us now! The Mercedes we drive, the Prada we tote! He could never have imagined the success we'd make of ourselves in Ireland. Look at us! Class is no longer an issue. Whether you come from Ballydehob or Ballymun, we don't care, as long as you have money. Townie or culchie? Forget about it, sure didn't J.P. McManus once drive a JCB for a living? In new Ireland there is nothing to hold us back. Yes, there's a recession on, but we've learned how to live like rich people now. And while we'll cut our cloth to fit new circumstances, we're never going back to the way it was before. No f**king way.

So in the boom years we got rich, bought some nice clothes and became a meritocracy; our horizons were limitless and there was nothing to hold us back. If this was the case, everyone would be wealthy, or at the very least, everybody would have a comfortable level of income. Ummm, that doesn't seem to have happened. The reason meritocracy is *supposed* to work is that if the old advantages of social class and status—townie, culchie or Anglo-Irish—are taken out of the equation, we are better off and will prosper more; if we all work hard and compete on the same level playing field, those who work hardest will benefit the most.

But there's another fly in the very expensive ointment. In a meritocracy, how do we cope with people who don't prosper or don't make it work for them? If we all have the same starting point, why do some people still end up poor? Are they lazy? Untalented? Were they sitting on their arses when things were good? Perhaps they don't deserve the same advantages as the rest of us. There is little room for passengers in the meritocratic society. If you don't make it, it's because you didn't try hard enough.

Is this what happened in the Ireland of recent times? Did the drive to become rich and profit while times were good overwhelm our ability to make good choices about how we lived, what we consumed, how we treated our environment and how well our communities worked?

The term 'meritocracy' first hit the streets in 1958 when Michael Young, one of the authors of the 1945 Labour manifesto, looked at the England he had had a role in creating and began to have second thoughts. In *The Rise of the Meritocracy*, he wrote about a society where 'if the rich and powerful were encouraged by the general culture to believe that they fully deserved all they had, how arrogant they could become, and, if they were convinced it was all for the common good, how ruthless in pursuing their own advantage.'[4] Doesn't sound that unfamiliar, does it?

Ireland isn't alone in this phenomenon. We only have to watch an episode of *The Apprentice* or MTV's *Cribs* to see how the code of living for the world at large is Work Hard, Get Rich and Tell Everyone About It (and if you live in a yurt in the Kalahari, get a satellite dish, for God's sake, darling). But while the race to be better or best worked for many, some people found it less than ideal and greeted the downturn in the economy with a secret sigh of relief. 'Thank Jesus!' many said to themselves while reading the belt-tightening economic forecasts. Here is a legitimate reason to put the brakes on our crazy levels of spending. Keeping up with the neighbours' yearly car changes and Caribbean holidays had all got a bit exhausting.

So, while several generations ago upward mobility was limited and the chance you might become the town millionaire so distant it was laughable, it became possible in the new Ireland, but it brought with it other unforeseen problems and anxieties. As the economic tide made lots of people, on the face of things, richer in Ireland, comparisons with those we live alongside also created anxiety over our own status versus that of others, and how well we were doing in relation to everybody else. Philosopher Alain de Botton terms this phenomenon 'status anxiety' and tells us that the rise in material wealth has 'gone hand in hand with a rise in levels of status anxiety among ordinary Western citizens, by which is meant a rise in levels of concern about importance, achievement and income.'[5] Being stressed about how rich or successful you are in relation to those around you is no way to live, but in the Ireland of the boom such comparisons were inevitable.

For many of us, this had become a way of life in itself. If you're a high earner and a busy person with no time, you'll be happy to know that some researchers have classed your woes as nothing more than 'yuppie kvetch' or 'complaining'. Complaining and moaning are pretty popular activities in Ireland as it is, so it's not surprising that complaining about being *too* economically productive and busy was something that would take on a life of its own. In the boom years, being busy and overworked became a strange badge of honour. Being busy was seen as a positive, privileged position, as it is generally high-status people who work long hours and feel busy.

THE CHIP AND PIN PARTY

But there's a problem. Busy people don't stand at their gate and chat to the neighbours. They don't tend to know who their local priest is ('Priest, what's that?') and are unable to identify people who live in their area who share situations and difficulties common to theirs. They could happily live next door to a dead person for days and not know it. Busy people don't tend to know much about the food they buy and put into their mouths. They don't have time to think about community, environment or the other things that the *Irish Times* tells them they should worry about. Lots of us are busy people, but did we want to be? Is this a good thing?

As we got richer in Ireland we became all these things: busier, time stressed and less occupied with what was going on around us. We moved from culchie habits, such as chatting outside the church, to glamour habits like throwing huge weddings and kiddies' birthday parties to impress others, driving giant German cars and shopping in New York. Ironically, the richer we became and the higher up the ladder we climbed, the busier we were and the more conflict this created in our lives. A paper published in 2008 found that work–life conflict is higher among professionals than non-professionals.[6] It seems that the more skilled and the less culchie you became, the more stressed and time poor you ended up. This is a sentiment frequently heard among women who find that as they get more senior roles in the workplace, the time allowed to spend with their children, family or at leisure shrinks. ('Leisure, what the hell is that?')

But even if the chip and pin party is now over, it's still the case that for ten years or so we embraced it with open arms, and probably will do again, once we've cooled down the red-hot credit cards for a bit. But are we happy to embrace this kind of lifestyle as the code we live by, and if we do, what makes us in Ireland different from everybody else? Is this how we really want to define ourselves? As a bunch of overworked consumers and shoppers? At the end of the day, are we Irish just the same as a group of people in Massachusetts, in Manchester, in Marseille?

No, not at all, we say. We're Irish! We're totally different from everybody else! We're Paddies, we're great fun, we're social, we love the chat. We have beautiful countryside, and Guinness, and we make lots of dairy produce and beef. We're the Emerald Isle, aren't we?

One function of knowing who you are is identifying those things that you are not; things that are alien, 'other', opposite to yourself. In the past these differences were often based on religious, class or geographical difference: the rivalries between parishes or towns marked how strongly connected people felt to where they came from. These feelings of 'sticking with your own' were usually expressed through sport; after all, what better way to get rid of local jealousies and anxieties than beating the crap out of one another on the football pitch? Until, that is, the Travellers landed in town and the whole lot sided together against a bigger common enemy.

IS IT BECAUSE I IS BLACK?

Having villains allows us to define particular people as bad, and us, by default, as good. Developing decent enemies makes your own community feel stronger. For years in Ireland, aside from our well-worn adversary, Britain, enemies were also local or geographical forces; the next-door parish that offered no help when your village was flooded, that townland that took the opposite side in the Civil War and grassed on your grand uncle. In the North it was easy: you stuck with your own community, swords drawn against those on the other side. While some people chose not to take sides in the North, they would still admit more closeness to one community than to the other. All of us like to feel we belong somewhere, whether it's supporting our local team or protesting against the landfill site planned in our area. By gathering together and sharing a common identity, we feel safer against threats from the outside. But are these allegiances changing, and are we in Ireland now less reliant on community and having a shared identity?

And what about our new multicultural community? There's no doubt that the speed with which some of these new populations arrived meant that many people found their presence confusing, threatening or unwanted. Some reacted to the influx of foreigners as to an invasion of the body snatchers: the newcomers would take our jobs, our livelihoods, they would steal our children and roast them on a spit over an open fire. There was no doubt that on occasion, the sudden appearance of large groups of asylum seekers or immigrant workers took people by surprise, and produced strong and often defensive reactions in rural areas and towns such as Clonakilty and Ennis. Several years into the journey, we can see that attitudes towards newcomers are still very varied. Most people will accept the necessity of Polish farm workers or Latvian chambermaids, but a more unwelcome tone is often directed towards African asylum seekers.

In Ireland the politicisation of religious and cultural origin is something that, hopefully, is receding into the background over time. The North is looking better; the high economic tide of the last ten years floated boats up there as well as down here. After all, Belfast had IKEA first, they should be happy, shouldn't they? The politicising of space and landscape is still evident in West Belfast's painted kerbstones, but now they draw tourists on special Troubles Tours. While the Downing Street Declaration of 1993 and the Good Friday Agreement have resulted in lessening sectarian tensions, identity is a hard-won thing and many still refuse to give up the old myths and tracts in order to emerge with new ones. To be a politicised Loyalist or Republican is to know who you are, and that in itself is a thing to keep hold of in an Ireland increasingly globalised and taking on board changing identities and meanings all the time.

So who are we now? If Ireland is a multicultural, increasingly urban state, where does this leave rural, culchie Ireland? For many people, if you came from a rural area it meant that generations of your family had come from the same place. You felt rooted there, and if you were involved with agriculture or the farming sector, your ties with the physical landscape and the community of people who lived there were likely to be stronger. If Irish people in general are under pressure to have what used to be considered 'urban' and 'cosmopolitan' traits—to be gym fit, fashionable, health conscious and conspicuously wealthy, what'll happen to the packet of Tayto and *The Sunday Game*?

FAKE BAKE

What about our media? Is it what we watch and read that has made us less culchie, less Irish? While our grannies bemoaned the overt sexuality of Elvis Presley's tight trousers, parents today groan inwardly at the sight of their youngsters glued to MTV shows in which teenage children squeal 'I hate you, Mom!' when their birthday present Porsche doesn't arrive in the colour they want.

We are told that losing weight, changing our lifestyle or undergoing plastic surgery will make us feel more youthful, sexy and successful. 'Transformation' is the new TV buzz word: clearly our lives, bodies, homes, gardens, eating habits and the way we rear our children all need improvement. Watching these programmes would discourage most Irish people from going out the front door; it's broadcasting that trades on the fear factor of not being good enough, successful enough, rich enough, thin enough. Okay, so we're not just a bunch of shoppers then, we're a bunch of badly dressed, fat shoppers who need surgery.

While American dramas and formats dominate our television channels, we still have the great white hope of local radio, TV stations and newspapers to offer a product to those who want to stay more local in the news they consume. (After all, you won't find Sky News reading out death notices.) However, the dominant trend is still towards a more homogeneous media experience. Spin FM is a successful Dublin radio franchise that has expanded into three regional areas. The publicity blurbs launching the new stations would make you believe that these are stand-alone stations offering a local product to their local target youth audience. In an interview with the new Waterford franchise after its inception in 2008, the station manager excitedly sold us a station that is in essence the sound of the south-east. In fact, it's very much not. Spin is the Supermac's of Irish youth radio; a 'one size fits all' model, the only regional difference being the accents of some of the presenters and of course the all-important advertisers. Spin is known for the high quotient of 'urban' music that it plays, basically music of black origin: R&B and rap. It's a slickly packaged product, but to make out that this is in some way tailored especially for regional audiences in Ireland is like pretending that Rihanna was born in Bantry and not Barbados. Of course, under-25s in rural areas want to listen to Rihanna, and the Artic Monkeys too, but telling rural Ireland that this generic stuff with a commercial heart is produced for their needs is simply disingenuous.

American and UK culture have always influenced Ireland, but it's the extent to which they act to homogenise us or threaten our own culture that prompts people to act. Those in the Gaeltacht regions would say that the Irish language was under the shadow of the guillotine a long time before television entered our homes, but this particular erosion is no recent development. To track the beginnings of the decline in spoken Irish you would have to go back as far as Ireland's settlement by Anglo and Scots populations. Perhaps Irish-speaking areas are one example of communities battling to keep a large functioning part of their traditional culture: their language. But would they have stayed this way without financial intervention and artificial support? Leaving aside government funding, do we care about keeping the Gaeltachts alive?

As Ireland is continually in a state of cultural flux, have we simply been experiencing something that we have always experienced—changing identities—or has recent change taken place at a much faster rate? Where has our inner culchie gone to? Has it been lost in a sea of Fake Bake tan and Tommy Hilfiger sweatshirts? Or is it alive and kicking in the tribes of GAA fans who show us that county and origin are still important enough to risk getting smashed over the head with a hurl? Does where we come from still excite passion or pride in Ireland?

It's likely that, to subsequent generations, what townland you are from or who you might be connected to will matter less and less. Many of the social networks in which young Irish people now operate are online ones, where common interests, hobbies or musical tastes are what draws their social group together. Less important are the local parish, the rural traditions and the fact that your aunt's cousin is a daughter of the woman who owns the shop in Ballymoate. Will the Ireland of the future be a place where rural and urban experience melts together into one dominant force, a melange of American, media- and consumer-driven culture that is shared by millions of people across the world, and not specific to any place in Ireland?

ANGELINA JOLIE'S ARSE

Global celebrity culture has proved that many of us have more interest in the relationships, reproductive lives and arse size of people who live in a suburb of Los Angeles than we do in someone who lives down the road. Because we gossip and share information about them, celebrities have become our new local icons. Instead of trying to ape Mrs O'Neill, the well-dressed bank manager's wife from up the town, we now fancy the look and lifestyles of people who live on the other side of the world. Celebrities function as the glamorous and inaccessible people we watch from afar and aspire to be. Those with extreme wealth, power or both have always excited interest; in the past they might have been very large landowners, local politicians or the first family in town to buy a television. The difference now is that the people we ape tend to be not local but international. If Angelina Jolie grew up and worked in a pub in Sligo, she'd probably hold little interest for us.

So where has culchie Ireland actually gone? Perhaps it's alive and well but we've simply hidden it, dressing ourselves in a different suit of clothes and losing our local accents. For a while many in Ireland pretended to be something else. It was deeply uncool to know anything about silage, *The Sunday Game* or who's in charge of the group water scheme. Young women from the country working in Dublin remade themselves into Carrie Bradshaws. Their talk was of *Cosmo* and noodle bars. They wouldn't dare let on that they spent the weekend helping their da dip sheep.

Yet we also know that while MTV may be blaring in the living room, there are kitchens all around the country where Muircheartaigh's frenzied GAA commentary blares from a radio with its dial so dust-gummed into place that RTÉ is a permanent fixture. Those who travel home from Dublin on a weekend train still return to houses where the local paper is read cover to cover, even the ads for rotor-cut balers—after all, that might have something to do with someone you know. In culchie Ireland, change is slower. Big dinner still happens in the middle of the day, there are floury spuds on the table and someone is very likely to come in the door for their tea at four o'clock, rub their hands together and say, 'Well, do ye know who's dead?'

Last year, writer Pat Fitzpatrick argued that culchies were second-class citizens in present-day Ireland. His view is that Dublin has forgotten where it came from.

It seems as if Dublin has divided the rest of Ireland into three parts—there is its own catchment area, the commuter counties around the east coast; then there is 'Cool Culchie Ireland'. This is made up of a few select villages on the west and south coast such as Roundstone, Kenmare and Kinsale, which are effectively Sandymount with tractors; and, finally, there is the rest of the country. The midlands and mid-west, a few mangy cities such as Limerick and Cork, that might occupy the same piece of rock as Dublin, but can't seriously expect to be in the same league.[7]

But perhaps the reverse is true, that rural Ireland has drifted into urbanised Ireland.

It's ridiculous to think that rural Ireland is full of farmers when there are only 40,000 full-time farmers left. In fact, it's full of people doing what people do everywhere else, but in addition, rural people also go to shopping malls like Liffey Valley and Dundrum, where many of the accents you can hear on any given day are clearly from far outside the city. It may be the case that we are all culchies at heart, but in the gloss of the Celtic Tiger we hid our identity in order to become the Carrie Bradshaws of our small island: worldly, fashionable and careful not to be defined as from anywhere in case it defined us.

There have been and there will be many different Irelands, but it's important that whatever identity we choose to be is one that is good for us, and one that offers individuals and communities a quality of life that suits everyone, not a culture in which the wealth and status of others overshadow how most people live. Perhaps being a nation of culchies was not such a bad thing. After all, knowing about the countryside and growing food is something we should be proud of, not something we should try to hide. It's becoming harder to find the things that make us unique in Ireland, and if being a green, rural nation was the photograph of our past, could it still be the photograph of our future?

Chapter 3

⎮ THE WAY WE ATE

Suzanne takes a trip down Lemon Meringue Lane

In the 1970s my parents were always going to Functions. I still don't really know what a Function is. I think it meant a night out, eating and drinking at the golf club. Or perhaps a dinner dance or some kind of school fundraiser. For us kids these were happy occasions; we got to stay up late with the babysitter and watch the television close down into a little white dot. Rules went out the window and sometimes we even got to watch *Tales of the Unexpected* on UTV, which sent all of us to bed with a new array of nightmare-spawning material that you could never tell your parents how you came across in the first place. Everybody was a winner.

My mother went to a lot of trouble to prepare for these evenings. From dinner time onwards, the roar of the hairdryer drowned out all sound within a two-mile radius. Fingers were burned on curling tongs and much time was spent in front of the mirror, a Silk Cut between two fingers, applying green eye shadow and parading to and fro in a selection of long polyester dresses à la Alison Steadman in *Abigail's Party*.

As well as whirling round the dance floor and being a bit of a glamour puss, my mother also liked to entertain at home. This was the era of the prawn cocktail, chicken Maryland and the sticky star of the show—lemon meringue pie. Other favourites that produced squeals of excitement in our home were Black Forest gateau, baked Alaska (usually collapsed from us sneaking open the oven door) and wobbly jelly turned out of moulds with fruit suspended in the middle. Food was heavily coloured, heavy on quantity, and garish luxury was the running theme. No big occasion was complete without a bottle of Mateus Rosé on the table or the venerable Blue Nun, a wine that sold six million bottles a year during the 1970s. The other star of the show was its

Liebfraumilch rival, Black Tower. Curiously (I'm being kind here), it's still Germany's number-one exported branded wine.

No surprise, then, that currently there is a revival of our much-loved (or hated) 1970s foods. Celebrity chef Heston Blumenthal included a Black Forest gateau as one of his prized dishes on the recent BBC series *In Search of Perfection*, and at Marks & Spencer there's been an extraordinary run on 1970s desserts, with sales of cheesecake, tiramisu, Black Forest gateau, profiteroles and apple strudel all up by almost a fifth in 2007 and 2008.[1]

I imagine much of this popularity probably stems from people in their thirties and forties harking back to when they grew up. It's a bit like listening to Red Hurley again or playing Kerplunk on the sitting room floor and sticking the spikes into a nearby sibling. My most memorable moment of 1970s food nostalgia was probably the day my parents came home from a supper dance and described how dinner had been the enormously novel presentation of scampi and chips served in a basket. A basket! With a red and white gingham napkin interior! This was far too much glamour for my eight-year-old mind to cope with—I think I nearly passed out there and then.

Photographs of our birthday parties in the 1970s reveal gigantic spreads of cold foods, trays of fairy cakes sprinkled with hundreds and thousands, and colourful plastic swords spiked through cheese and grapes or columns of ham and pineapple. The overall effect was of a Pucci-like display of overloaded colour; no wonder several photos of the dinner table had to be taken before the bunch of wide-eyed, ravenous children were let loose to demolish it.

This was 1970s eating at its best. Ireland was at this point waking up to food as having other, more glamorous, possibilities than plain old day-to-day sustenance. This was the advent of a new age as food moved from being the sustainer of life to being something more playful, something that could provide such alien possibilities as enjoyment in preparation, enjoyment in eating and, from time to time, being a flamboyant indicator of status. As food writer and retired chef Gerry Galvin points out, 'The whole idea of eating for pleasure was not acceptable in Ireland until very recently.' For many, the idea of the famine was hovering in our distant memory, and food was something to be grateful for. Eating for pleasure alone, and revelling in preparing food, was something that Irish people would not engage with until the latter half of the twentieth century.

This was the time when Irish cooking—the fabulous rural meals prepared in farmhouses all around the country—was seen as staple fare not fit for a restaurant dinner table. It was assumed that *proper* food was imported or was of French origin, and dishes such as the flambéed Steak Dianne or Boeuf à la Bourguignonne became stars of 1970s urban cuisine. Thus Ireland's slow awakening to the possibilities of food began with classical French, and our first celebrity chefs and kitchens, such as Sean Kinsella's Mirabeau, served these up to eager 1970s diners. The Mirabeau, a favourite of Haughey's, notoriously had no prices on its menu, a bold statement at the time but one which served Kinsella well, giving the restaurant the rarefied allure of a top Paris kitchen. In 1981, the Gauls themselves arrived, with Patrick Guilbaud bringing his blend of elaborate French cooking to his first location off Baggot Street. Eventually his craft was to earn him two Michelin stars. Others followed: Arbutus Lodge in Cork, an Irish-French restaurant run by Declan Ryan and his wife, Patsy, won a star in 1974 and Roscoff in Belfast gave Northern Ireland its first Michelin-rated restaurant in 1991.

A NICE BIT OF TONGUE

For most people, reading about the goings-on at the Mirabeau or Guilbaud's in the Sunday papers was as close as they got to stepping into Ireland's gourmet world. Restaurants were only for special occasions, and in most parts of the country were seldom visited at all. Most food—plain combinations of meat, vegetables and potatoes—was prepared and eaten in the family home: nipping down to the Spar for pizza and a bottle of Merlot was light years away. Processed foods were scarce and traditional Irish eating habits prevailed. Many homes still ate fish on a Friday, giving fish a lowly place on the food ladder. For most housewives a supermarket shop involved sticking to a firm budget: in 1973, families spent a whopping thirty-two per cent of their total income on food. A week's worth of dinners entailed rigorous economising, getting the most out of what you purchased and never, *ever* throwing anything away. Most women's understanding of food, how to prepare and cook dishes such as tongue, tripe and kidneys, was routine stuff. Offal and cuts of meat that nowadays send people scurrying under the table were regular features on the family menu. Everything was for eating: whether or not it still had pretty eyelashes on it, using every available part of an animal's body and any type of vegetable you could get your hands on was the way we ate.

The memory of peering into a large saucepan full of boiling tripe is not what you'd call fantasy eating, for alongside the 1970s psychedelic delights, our household was still a mecca for all the traditional stuff: big chunky oxtails, glaring cod, aggressive pinching crabs whom I cried over in their boiling tombs of salted water. My parents were firmly of the old school, and despite the fact that my mother worked, baking was done in our household most days of the week. I never thought it unusual that all our bread came from our kitchen oven or that the garden produced our potatoes, carrots, summer vegetables, berries and apples. Though by the time they were in their thirties my parents were living a suburban life in Wicklow, they probably didn't know the extent to which they were still attached to growing their own food and how hard it was to shake off old ways of life. Both were from farming families in Donegal and Fermanagh, their own mothers the type of scary women who could calve a heifer, bake a Victoria sponge and knit an Aran sweater for their ICA evening all at the same time.[2] It was unthinkable in their family homes not to be able to kill and prepare stock for the table and, more importantly, to have ridiculous amounts of cakes, fresh soda bread, homemade preserves and honey from your own bees available at any time of the day should someone drop in for a pot of tea so strong you could trot a donkey across it.

And God love the poor visitor who wasn't prepared for the Huge Food Challenge of visiting my grandmothers' rural homes. Once a car was spotted approaching up the lane, a table groaning with food miraculously appeared in ninety seconds flat. By the time the guest arrived, an effortless display of cold meats, cheeses, pickles, breads, salads and an array of cakes would have appeared as if by Photoshop, by which time granny had changed into her best cardigan and pearl brooch, apparently having barely stirred from her knitting. For the guest, the price of this enormously calorific hospitality was that they had to eat it. *All* of it. After all, there could be a famine again, at any moment! No one was allowed to creep away from the table unless suffering an actual cardiac incident. Having cleared all plates, you might exhaustedly bargain a departure if you promised, yes, you would take home a dozen eggs, five turnips, two nasturtiums and a pup. There was always a pup.

For my grannies, a key part of the experience was feigning surprise at how their cooking could be anything more than ordinary. They pooh-poohed any thanks, waving an indifferent veined hand and saying, sure what else did they have to do all day? From time to time the act would slip when a valued guest (usually some roving man of the cloth) heaped praise on a particularly moist brack, or tomatoes that had a redness and richness of flavour far better than any they had tasted. Then you might just glimpse a moment of competitive victory before, clucking with pleasure, they headed back into the kitchen to make yet more jam, or shape up a few more trays of scones for the oven.

NINTENDO FARM

It's hard to imagine the behind-the-scenes exhaustion levels of running such food-intensive households when my parents were growing up. Activities began at dawn, when the hens and livestock were fed and the range filled with turf and lit. My grandmothers baked in the morning, so that the afternoons were left for sewing, mending, washing and gardening. My mother recalls taking turns with her siblings at the heavy and hated task of butter making, plunging milk in the butter churn, a degree of manual labour unthinkable for modern children. Other delights of her childhood were cows kicking you in the face while you milked them, bullocks standing on your toes while you screamed blue murder, and the arduous task of hauling up water from the dark, threatening well. (Falling in was not an option.) While many memories of life on post-war Irish farms are happy ones, it was an existence involving intense physical labour and an element of risk. There was always the chance that damp weather would blight the potato crop, or that if that godforsaken rain didn't stop you'd have no hay for animal fodder next winter.

No wonder our parents think that any kid born in the last forty years has it ridiculously easy. As late as the 1950s, children in Ireland were an important source of work in the agricultural economy. It was an acceptable and necessary fact that all hands, even little ones, needed to be employed in getting enough food on the table. In one onlooker's eyes, Irish girls in particular had a rough time of it. Child labour was child labour, but for girls, little status came with the job.

> Children begin to help in the house at an early age, drawing water seems to be the chief occupation of boys. As girls grow older, they share a great many of the household chores or look after younger children. When a daughter reaches sixteen, if she remains on the farm she must do a full day's work, and too often her life is one of unrelieved drudgery. There is an almost oriental attitude to girls. They are favoured by neither father nor mother and accepted only on sufferance. This is, perhaps, too strong a conclusion, and it would be better to say that they are loved but not thought of any great importance.

So wrote McNabb in his *Limerick Rural Survey*,[3] undertaken in the uncomfortably recent period of 1958 to 1964.

If you were unfortunate enough to be an illegitimate child who was in the care of the local authorities, the chances were that you could end up being hired out by your guardians as labour, a practice documented as happening in Clare as late as 1930, when 112 children were hired out as farm labour and domestic staff.[4] Most hiring seasons ran from St Patrick's Day to the end of November. Even farming families who employed hired labour expected children to do routine work on the farm. Being close to food production taught you exactly how valuable the stuff actually is, and provided a link to the basics of life itself. The growing of food is something far distant now to most of us, though still present in most of the less developed world, and it has only gone from many Irish homes in the past thirty years or so.

Never having had the most exciting diet in the world, Ireland was still very dependent on the potato, a trend that has only altered in the last ten years when sales of pasta, rice and other carbohydrate-rich foodstuffs have overtaken the common tuber. So unpopular is the potato currently that growers are encouraging its use through such schemes as the Year of the Potato and awareness campaigns highlighting its nutritional value. And certainly, the potato has been a good servant to Ireland. From the sixteenth century, much of the country survived on a diet of potatoes and milk and took their meat or fish when they could get it: herring if you lived near the coast; poultry, offal, and cheap cuts of lamb or beef if you could afford them. For generations, Irish eating habits were simply about having *enough*. For farming families or those with enough room at home to keep pigs or poultry, feeding yourself by growing your own produce was a natural process.

This was a time when food held no particular magic and wasn't particularly complicated—it just had to be plentiful, well prepared, and eaten. Regulations on food safety and food preparation were non-existent. In that time, a butcher could take pigs from your farm, butcher them and sell you back the sausages from his own shop. Traceability didn't exist as an issue because most foods came from local sources, or your own back garden produced the chicken that went into your pot. The skills of butchering animals and growing vegetables, and the basics of feeding yourself, were learned by children from their parents. If skills only take a generation or two to die out, the knowledge of our parents and grandparents who had mixed farms and a variety of foodstuffs under production is already, in many cases, lost.

Knowledge of cooking was passed on within the family home: my mother and her sisters learned to cook from their mother, and so on. For those who needed a bit of help there were the books of Mrs Beeton and Fanny Craddock, strict matronly types that make an average school home economics teacher look like a drug-addled bohemian. For the more continentally minded, the books of Elizabeth David first appeared in the 1950s. Her first publication, *Mediterranean Food*, was shockingly new, describing unfamiliar ingredients, spices and the enticing dishes of the Levant gleaned from the years Elizabeth spent working in Egypt during the war. Her writing was decades before its time, her focus firmly on well-sourced ingredients and artisan cooking. Ironically, she described what was still taking place in kitchens across much of Ireland and marked out what is so popular today in terms of what we want from food. She described:

> … the bright vegetables, the basil, the lemons, the apricots, the rice with lamb and currants and pine nuts, the ripe green figs, the white ewe's milk cheeses of Greece, the thick aromatic Turkish coffee, the herb-scented kebabs, the honey and yoghurt for breakfast, the rose-petal jam ….[5]

Elizabeth David's introduction to the food of France came during a period spent boarding in a strict Parisian household. Reading of her time there reminded me of my own entrance to the world of French cuisine when, as a teenager, I au paired in a chateau close to Clermont-Ferrand. Sitting at one end of an immense table, I passed many bizarre evenings dining with only an ageing marquis for company, whose poor grasp of English, combined with my even poorer grasp of French, made for circuitous, Beckett-like conversations. It didn't help that our view of each other was generally obscured by candelabras and large flower arrangements, but we plodded valiantly on, each knowing that the other was completely at sea, eyeing progressive courses prepared by the chateau's chef, the likes of which were completely beyond the experience of any Irish teenager of the 1980s.

But after several months cycling through fields of sunflowers and swigging cocktails on the lawn with members of L'Académie Française, I became as blasé about good food as the rest of the household, browsing through the silver buffet dishes on offer at breakfast, expecting a decent glass of wine with lunch and failing to think twice about the poached pike, bacon cooked with green Puy lentils or other curious oddities on offer at dinner. When I told Madame La Marquise that pike is thrown back into the rivers in Ireland, she had an episode of mild fainting, for the fish was one of the most valued foods from the Auvergne, and we ate the best of them. In Paris, at the family's main residence, I grew to expect such dishes as artichokes dipped in aioli and *cuisses de grenouille*, which we enjoyed in between attending couture shows and sitting in a little gold chair à la Anna Wintour.

It was only in later years that I realised the unique opportunity I'd had to live and eat in a household of that kind. They were a family who were particular about food, not because of their rank in society, but simply because they were French. I later found that in all French households, regardless of income, the attention to detail paid to food is of the highest order. Their approach is that food can be simple and will always taste good if it is local, fresh and of the best quality. In France, no matter how busy people are, mealtimes are a time to slow down and use food to mark a social point in the day. Feeding people is not something to be executed quickly. After all, food is life itself. Far ahead of her time, Elizabeth David identified the key points of how we are told to rate food today: find quality ingredients, in season, and prepare them simply. Ms David understood the call of a nice bottle of wine, crusty bread and a sunny day. No wonder her recipes are still so popular today.

BANANARAMA

For most of us in Ireland, Mediterranean food didn't make an appearance on our tables until the late 1980s. Moussaka made its first curious debut in our house around this time and, following a visit to Morocco (maybe my mother was a hippy after all), we had the joys of couscous and, hot on its heels, lasagne and spaghetti Bolognese. Instead of Functions, my parents began going to the mysteriously named American Tea Parties. As far as I remember, these comprised of yet more eating and drinking, the difference being that the invitees supplied all the food, dividing up the menu between them to bring along starters, mains, salads and desserts. This was the decade of Cajun chicken, quiche Lorraine, cheesecake and, yes, the hissing, spitting Soda Stream. The fact that we can still experience dreamy-eyed nostalgia over a Soda Stream is indeed strange, though not as strange as an American I know who recalls his teenage 1980s food staples as sushi, cocaine and ludes. Gosh, I definitely grew up in the wrong country. We thought having a Teasmade was racy.

The eighties was when our eating and shopping habits of today really began. Home-made foods were suddenly unfashionable; instead, Ireland was sold the highly processed attractions of Easi-Singles, French bread pizzas, Angel Delight and Artic roll. Part of the huge rise in ready-made foods came from the advent of the microwave, a piece of equipment they told us would revolutionise our lives. Melted plastic containers aside, the microwave was a huge hit in Irish homes, despite the fact that the early models were the size and price of a small family car. In our house, the arrival of our brown and beige Sharp microwave was viewed in the manner of the second coming of Christ. *No one*, the parents threatened in scary voices, was allowed to touch it, at least not until the manual had been read from start to finish, and a few times again in case they'd missed a bit. As it squatted on the kitchen counter glowing and humming, we regarded our microwave with a kind of awestricken, mesmerised terror; a golden ark of the covenant, coming to rescue us from horrible home-made healthy foods and deliver us into TV dinner heaven. Delighted, I foresaw the months ahead as one long yellow brick road of crinkly microwave chips like on the telly, and if I fancied a change, melted cheese on potato waffles. The microwave was going to be my childhood saviour, an escape from Irish stews rumbling on the hob and evil vats of boiling cabbage from the garden.

Delivering fast, instant heat, the coming of the microwave spawned the age of ready-made meals. The new cook-and-chill technology meant that busy, time-poor families and people living alone could now buy seven plastic-coated dinners for the week and heat them up whenever they had three minutes and thirty seconds to spare. As more women worked outside the home, it meant that even little Tanya could cook her very own plastic rectangle and eat it in front of *Bosco* on the sitting room floor. At the time, this was what we were told would be how the modern family would function: individual meals, cooked individually and eaten at different times. It's a picture that sends food advocates of the present day into a paroxysm of despair. The eighties also saw the beginning of the rise in people living alone and the reduction in size of the average household. We watched the Gold Blend singletons flirting with a new way of life; owning their own homes, cooking for themselves from the space age plastic container. The ease! The convenience! How we loved the microwave lasagne's searing hot béchamel that ripped off the roof of your mouth, while your fork, exploring inside, hit the ice-cold layers of lumpen pasta beneath.

For most families in Ireland, the cult of the microwave was a slower grower than it was in Britain. With strong rural ties and a modicum of good sense, many Irish housewives made a sceptical face and stuck to preparing their own food. Microwave dinners were, after all, very expensive relative to what you could make at home. Ironically, during this time ready-made meals were aspirational foods, the stuff that rich families ate. The good food growing in the fields around you was far too unfashionable, despite being a thousand times better for you. Like McDonald's, which opened their first Irish restaurant in 1977, fast food and ready-made meals were far too expensive to be everyday staples. The thought that they mightn't actually be very *good* for you was a long way away.

In these postmodern times, microwave dinners have gone the same way as the New Romantics: a guilty habit to be enjoyed as rarely as your Human League records. But the 1980s changed our attitude to everyday food in a profound way. The vast array of frozen and imported foods meant our diet was not confined to what we produced in Ireland: any dish from around the world, whether a Big Mac or a curry, was now accessible to us thanks to the vast supermarkets whose dazzling array of products sold us food dreams from every part of the globe. But it wasn't just the ranges that supermarkets carried that were growing, they themselves were becoming bigger (and fewer) as they built new stores and bought up smaller companies, centralising the purchasing, distribution and often the manufacture of foods. The arrival of a supermarket in your town meant that many small outlets such as local butchers and vegetable shops closed. It was to be many years before, despite their convenience, supermarkets came to be viewed not as protecting consumers' interests, but as dictating how our food is produced and retailed, and ultimately as limiting our choice of what we eat.

KILLING ME SOFTLY

The 1990s in Ireland saw many trends in food from Britain and America really take root: supermarkets, fast food outlets and even garages were increasingly taking the job of preparing food away from the individual household. In 1980, the average meal took one hour to prepare. By 1999, according to UK figures, it took twenty minutes.[6] By 1999, chilled ready-prepared meals overtook demand for frozen ones. Surveys suggested that consumers saw chilled meals as better quality, with a greater range of recipes. The products provided meals that many people could not cook themselves. If food was getting more primped, preened and packaged, it was also getting cheaper. By 1994, we in Ireland were spending 22.7 per cent of our income on food, down from over thirty per cent in the 1970s.

But there was a strange dichotomy at play. Despite the continued rise of the ready meal, cooking was becoming a surprisingly popular leisure interest. At the same time that supermarkets were stocking an ever broader range of processed foods and ready meals, they also widened their range of ingredients, providing the makings of Chinese and Indian staple dishes, exotic fruits and vegetables, ingredients for sushi, etc., which reflected our new interest in cooking. Travel had broadened our Irish tastes, but while the idea of making sushi might have thrilled us in the supermarket, the reality was more likely to be that we went home and stuck an M&S chicken tikka masala into the microwave while watching Delia cooking spaghetti alle vongole on the telly.

In the 1990s, cooking became part of the new vogue of nesting. Television channels were crammed with shows related to domestic life: redecorating your home; redesigning your garden; cooking for friends. Cooking began to be seen as aspirational, something that was a skilled accomplishment rather than the boring drudgery that your mother moaned about. Men began to cook in their own homes—yes, even in Ireland. Jamie Oliver wrote that his approach to cooking originated from his struggle to recreate restaurant dishes in the cramped kitchens of rented flats, and his TV programmes and books tapped into a desire among ordinary people to create delicious and innovative dishes at home with a minimum of fuss. And let's not forget the added attractions of a bouncy young man flinging things casually into a bowl to create a dish of remarkable flavour and complexity. Who wouldn't want to mimic that?

The 1990s saw more Irish people travel to exotic locations and even buy second homes in places like France and Spain. Our appetites matched our aspirations and supermarkets filled the gap, creating the age of pre-washed, ready-prepared salad leaves, radicchio, rocket and lollo rosso, and the 'vine' tomato. Things like capers, olives and sun-dried tomatoes became familiar staples of Irish bistros and dinner menus. 'What the f**k is a caper?' I nearly blurted out at a smart Dublin household in early 1990. Clearly, *A Year in Provence* had become *A Weekend in Ranelagh*. Somehow we all believed that if we ate like Mediterranean people we'd no longer be muck savages from Ireland but Armani-clad beauties wafting around St Tropez without a hint of sunburn or the smell of slurry about us. We were always fond of a good fantasy.

While we got snobby about food, we also got shoddy. Our eating habits in the 1990s were changing fast, with more, far more, food eaten as snacks and on the move. Fast food outlets multiplied and running around with a takeaway coffee in your hand or stuffing your face with a sausage roll while driving to a meeting in Portlaoise changed from something that you had to do now and again into a normal enough day's eating. Sales of cereal bars, portable salad bowls and yoghurts with a spoon taped onto the side rocketed. I spent the late nineties working on RTÉ's farming and food programme *Ear to the Ground*, and ironically, I've never eaten so much junk food in my life.

Looking back, we had the typical time-poor, cash-rich food intake. We drove thousands of miles a month visiting farms and food producers all over Ireland, and as we drove, we ate. Breakfast rolls, King's crisps, hash browns, Burger King, sausage rolls, Danish pastries and sandwiches from garages. Wispas, Marathons, muffins, Bombay mix, bottles of Coke and takeaway coffees. Yes, we burned most of it off dragging camera gear through mucky fields, but at 1200 calories a pop, a couple of breakfast rolls could power the Kish lighthouse. We were typical of many people in Ireland at the time: increasingly knowledgeable about food and what fast food was doing to us, but reluctant to spend a lot of time cooking. So we went to the pub or headed home to watch Nigella instead.

It was also only in the 1990s that Ireland realised with shock that we used to make pretty decent food ourselves. As obesity was being spoken about in hushed tones and awareness grew that our expanding junk food habit might wipe us all out, a slow awareness of simpler, more flavoursome food choices began to appear. Since the late eighties, Darina Allen's stable of *Simply Delicious* cookery books had been telling people in Ireland that good food was perhaps closer than they thought. Her mother-in-law, Myrtle Allen, had as far back as 1964 opened a portion of her house in East Cork as a restaurant, with a food ethos of a return to simple local ingredients, a view echoed by the Slow Food Movement,[7] which advocates awareness of local food traditions and promotes knowledge of where food comes from, how it tastes and how our food choices affect the rest of the world.

Around Ireland, small food producers and the artisan food makers were finding a renewed interest in their products. Cheesemakers, bacon curers, berry growers all found themselves saying 'I told you so'. Our own Irish produce, produced locally and free from heavy processing and its retinue of additives, began to look very attractive to those seeking simpler, healthier choices. On *Ear to the Ground* we visited many of these hardworking and sometimes visionary food producers. Some of them were keeping alive skills that their parents had taught them, such as the lady in County Offaly who made preserves from quinces; some were larger outfits, such as the Grubb family whose Cashel Blue cheese has become an award-winning Irish food success. There are many such examples. These small producers had often moved from conventional farming to a more hands-on approach to food. Because they were more labour intensive and demanded quality ingredients, these products did not come cheap. Termed 'niche foods' during the 1990s, many are now moving up the popularity ladder as acceptable weekly choices in your shopping basket rather than occasional treats.

SASHIMI OR CRUBEENS?

It was also during the late 1990s and the glorious years of the yelping tiger that we all got richer (apparently) and began to eat out much more. The hotel and restaurant scene exploded across the country, with Hiltons, Mediterranean brasseries and sushi bars popping up everywhere, no longer aimed at the rotund American tourist but at *us*, the food-aware, cholesterol-conscious New Paddies. Lots of us became a bunch of food snobs. Destinations such as Dunbrody House, the Mustard Seed and Ballymaloe were packed out with foodie weekenders looking for something Irish, seasonal and 'that incredible lobster ravioli' to tell their dinner party friends about.

Capitalising on an era of loose money, many of the new Irish eateries served what was far from stellar food, but as dining out was key to the Celtic Tiger experience, we populated these places in droves, and often paid little attention to the origin or quality of the food on the plate. While new restaurants abounded, not all of them deserved to survive. Food editor Ross Golden-Bannon terms many of these boom-time eateries 'off the peg' restaurants: 'These are medium- to high-end places following a standard formula of buying in poor-quality produce and adding expensive garnishs. I'm amazed at how many people fall for it.'

While Ireland was said to be having its food moment, being seen out and about and eating in the right places was often more important than *what* you ate. Many diners were still preoccupied with the surface-level stuff and didn't have the interest or knowledge in looking at what was actually being served up to them. So on the one hand there looked to be an explosion of Irish 'foodies', though some clearly had more right to the title than others. 'There are two kinds of foody, those who have a palate and then there's the wealthy, many of whom don't have a palate but who were concerned about the status of where they are seen eating,' says Golden-Bannon.

Celtic Tiger restaurants saw a huge amount of the latter. In that period of plenty, getting your bum on the seat of a new restaurant before it had actually opened to everyone else was crucial to maintain your place on the food status ladder. Much-heralded openings such as Venu, Mint, The Saddle Room and Locks were anticipated with giddy excitement of middle-class diners who'd swapped cocaine and clubbing for kids and eating. Food was the new black, a point firmly brought home to me when a friend served a Raymond Blanc dish at a dinner party that took *thirteen hours* in total to prepare. High-end joints such as Thornton's and Guilbaud's were, in restaurant speak, out the door, with waiting lists of months or more for tables in the evening. Noughties hallmarks of the top-level kitchens were sea urchins, scallops, wasabi, foam, truffle shavings and a return to delicate nouvelle cuisine presentation. As cod raced up in price and scarcity, fish such as pollack and ling that had formally been seen as very much D list were dolled up and served as fashionable fare. For the 'civilians' there was always bresaola, Caesar salads, lattes and the lunchtime wrap with tapenade.

In the blink of an eye, Irish eaters had become discerning and demanding about their food. 'Organic' and 'Irish' were the buzz words that we looked out for on menus. Our food knowledge had improved, and we sniffed at the idea that anything that hadn't had a decent shot at life in a decent, clean environment might end up on our plate. At last, Irish food was recognised as fashionable, desirable and good to eat. In chef Kevin Dundon's book *Full on Irish*, the foods we regarded as old fashioned and unappealing were making starry comebacks: honeyed root vegetables with a red wine jus, petit pain squash and roasted garlic soup.

But in the face of leaner economic times, can this sort of luxury food stay viable on restaurant menus? Food writer A. A. Gill has written that 'one of the great bonuses of the imminent grand depression is that it looks as if the first casualty will be the utterly bogus and cynically manipulative organic food movement'. Stop reading now if you grow organics or spend half a year's wages on your organic shopping habit. He continues: 'The vertiginous ascent of costs has been a straightener for restaurants; apparently, cheap simple dishes have doubled in price. Using organic ingredients when there is no discernible taste or digestive benefit, only the mild sense of fashionable self-righteousness, is hardly worth it.'[8]

Many would beg to differ: most artisan-produced or organic foods do taste better than mass-produced, non-organic foods, but what is also clear is that they are still expensive and that organic produce comprises a tiny minority of what is farmed in Ireland. Alongside our boom in luxury eating, it was still convenience foods and cheap, processed supermarket offerings that filled many family fridges. At the same time, a new breed of retailers arrived from continental Europe: the discount chains, such as Aldi and Lidl, that offered basic goods in large warehouses at rock-bottom prices. The dilemma that people faced was that processed foods were quicker to prepare than fresh, and no matter how many Gordon Ramsay extravaganzas you watched, a pizza was quicker to cook from frozen or order in than to make from scratch. 'I call it the Curse of Cheap Food', says writer and critic Tom Doorley. 'As the proportion of what we spend on food has fallen over the years, food itself is less valued and less of a priority. And not just for those on a low income. You'll see people who spend a lot of money on their cars or their holidays still buying really cheap food that is mass produced and has little nutritional value.'9

Yet on TV we are being bombarded with information with what our lazy habits are doing to our health, programmes that endlessly tell us our convenience food diets are sending us to an early grave. Following Morgan Spurlock's film *Supersize Me*, many television and media experts drove home the same point: return to fresh vegetables and unprocessed foods or your insides will turn to sludge. Obesity in children and type 2 diabetes are on the rise, and while food is cheaper than ever, it appears that most of it is bad for us. Bad with a capital B. Turn away from the convenience food devil, we were told, return to the way our forebears lived.

Food and gardening programmes told us the path to salvation is to grow our own herbs, and if we can manage that, throw in a few seed potatoes and tomato and carrot seeds and see what happens. It turns out it's relatively easy to grow some of your own food, rear chickens and source organic or local produce. At a friend's farm I recently watched his rare breed pigs root busily through a paddock that they've turned into glorious mud. If you have the time and the space, producing your own livestock is rewarding and can make you a few bob, and I'll be the first to say the meat from these happy pigs tastes like stuff made in heaven. Wrestling your own rare breed pork to the ground is not for everyone, but strangely, doing stuff the hands-on way is all a bit of an ironic return to what my parents did at home. Grow your own, organically, and you'll save yourself money and a shedload of chemicals and additives.

THAT TRIP TO TESCO JUST COST ME €160

Our acceptance of food as being cheap and easily available took a buffet in 2007 when grain and rice prices first began to soar across the world. A combination of drought and speculation, the spike in the price of staple carbohydrates was soon passed on to the consumer. Oil prices sent farm inputs soaring, and as animal feed is largely made from grains, most things we ate suddenly became more expensive. It was a reality check on the degree to which the price and availability of food was out of our control.

For decades, food had been cheap, plentiful and something we took thoroughly for granted. Suddenly we began to sit up and take notice. How did we let the production of food get so far out of our own hands? It was a good time to think about who provides the majority of what we eat (multinational food companies) and what it was costing us in terms of health, sustainability and melting the ice that was keeping our giant gin and tonic lifestyle afloat. Maybe it was time to think more seriously about growing those peas. After all, it would be better than having them flown from Kenya in the middle of winter—they'd look better, taste better and there'd be the inevitable rush of pleasure at having produced something so wonderful yourself (nature has zilch to do with it, of course). Failing that, you could pick up some peas in a farmers' market or organic shop. Yes, it's more expensive, but examine your supermarket receipt next time you exit and you'd be surprised what they are charging for the simplest of produce.

Consumer agencies tell us to shop around to minimise our ever-rising food bills; go to local butchers and greengrocers, it's bound to be more locally produced and cheaper than what the monolith supermarkets supply. Check the origin of stuff on labels, eat less processed food and stick to fruit and vegetables in their growth season, when they are cheaper—yes, it might be nice to eat asparagus in winter, but flying it in is hardly ideal, for loads of reasons. The current vogue for allotments, self-sufficiency and home-grown vegetables is no different from what most families did fifty years ago. The reasons for returning to simpler methods of growing food are slightly different, but perhaps it's time to get the gardening gloves out and start. No matter how sophisticated we think we are in Ireland, we are still a food nation, and that closeness to agriculture in both our heritage and in practical terms may have some hidden lessons and advantages in a future where food could be more expensive and complicated than it's ever been before.

Even for a nation of sceptics, it looks like we're becoming more aware of this and our knowledge of food and nutrition is thankfully improving. The National Health and Lifestyle Survey published in 2008 found that, in general, we are becoming more health conscious and aware of what we are putting into our mouths. Half the population recorded self-rated health as 'excellent' or 'very good', an increase since the last survey in 2002.[10] Similarly, there was a reported increase in the number of respondents who described their quality of life as 'good' or 'very good'.

The demon drink, the rise of which seemed to follow the upwards trajectory of the Celtic Tiger, has also received a matronly rap on the knuckles; the number of respondents who consumed more than six or more drinks at least once a week (termed 'risky drinking') had fallen since 2002. Again, our present-day eating habits show a dichotomy in our behaviour. Half of respondents reported snacking between meals, most commonly on biscuits and cakes, and eighty-six per cent of us ate more than three daily servings of the foods from the top shelf of the food pyramid, where the bad stuff is—the fats, sugars and salt. Almost one-third of respondents either always or usually added salt to food while cooking (thirty per cent) or added salt to food at the table (thirty-two per cent). Overall, ten per cent of respondents had not eaten breakfast on the day before the survey. There's still a lot of garage grazing going on, with seven per cent purchasing their breakfast, twelve per cent their main meal and twenty-three per cent a light meal outside the home.

HUNT YOUR DINNER DOWN

We still have some bad habits—time and convenience are still huge drivers in how most of us shop, prepare and eat our food—but overall, we Paddies are definitely more food aware, and many of us suffer mild panic at how lazy and couch potato-shaped we have become. We think about how healthy it would be to grow our own food, but how many of us actually do it? The recent popularity of allotments, veg growing and using wild food has created a new vogue for man, and woman, as hunter-gatherer. Somewhere in a deep and totally unrealistic part of our brain, this notion is quite appealing. Yes! we think, I can grow my own veg, I can make my own bread, and if I get a shotgun and a licence I can even shoot wood pigeon, break off the wings and turn it inside out into two perfect barbecued breast portions, just like Gordon Ramsay did on *The F Word*.

This hunter-gatherer craziness mostly appeals to men who, let's face it, have little opportunity in the modern Western world to hunt or gather anything except women around bars. Perhaps it appeals to the side of men that is depressed or emasculated by another Monday morning at the office, in which case, even I'd admit that the fantasy life of a shotgun under your arm, bagging pheasant and spearing fish à la Bear Grylls is a pretty attractive one (even if it turned out the production crew caught it for him). Over on Channel Four, their *Wild Food* series gave us the cutesy pairing of Tommi Miers and Guy Grieve as they capered across England and Spain, shooting boar, skinning rabbits and hanging out in a tent looking like something from the VIP area at Glastonbury. Real men are harking back to real things, and that includes masculine ways of providing food. Even Irish chefs are at it; on *Guerrilla Gourmet* we had Kevin Thornton hunting rabbit with a couple of ferrets for a feast of old-style Irish food served in the Rock of Cashel. It seems that the early twenty-first-century food fashions of wild food and growing your own are the equivalent of our 1970s lemon meringue pie.

Food can be as much a victim to fashion as anything else, and it's interesting to speculate how recent trends for organic, sourcing local and growing your own will fare in the long term. One thing is certain: rising food prices and health scares arising from heavily processed foods will keep the emphasis for a while on eating simpler, seasonal local fare. And if we want it to be as local as our own gardens, we're going to have to put a bit of spadework in. After spending nearly fifty years living close to Dublin, my father still makes jam from his own fruit trees and feeds a family of goldfinches in his garden. Some rural habits are hard to leave behind. Perhaps if we think about how we lived in the past and take parts of it back into our own lives we might eat better, feel better and, as the Dalai Lama says, have time to smell the roses in the rose garden.

After all, good food comes from the simplest of things. The fact that most of what we eat is controlled and prepared by huge multinationals has made food remote and out of our control. We often feel confused and unknowledgeable about what goes into it and what is truly good for us. With less money available in Ireland, it would be a shame if we forget that the quality of the food we eat has important ramifications for our health, the environment and the wider economy. The food we produce in Ireland is highly regulated, great quality and provides local income and employment. Relying on mass-produced food produced far from our shores could turn out to be a very expensive mistake.

As Ross Golden-Bannon says. 'There are plenty of people in this country who say they are committed to good food sourcing, but for the majority price is still the deciding factor.'

So while money is tighter, it's up to us to decide how important good food is to our way of life, whether it's fashion or for real, whether food safety scares and the awareness that comes from them have a real impact on what we are prepared to feed ourselves and our families. By even making the occasional change in how and what we eat and by shopping more locally, we are getting a bit of the control back for ourselves. Food can be merely fuel, but it can also be pretty interesting stuff. It's also one sure thing in life that you're never going to escape from. If we are informed about food, make it enjoyable and think about what our parents ate, we will benefit our health, our environment, our economy and our taste buds along the way.

Chapter 4 ∾

| WHO IS RURAL IRELAND?

A farmer wants a wife

Martin is the wrong side of forty and looking for a mate. His mother circled an ad in the personals of the *Irish Farmers' Journal* and handed it to him last Friday when she came back from the town with the shopping. The ad was in the 'Women Seeking Men' section. It read:

> **Genuine and Respectable** I am a single attractive lady in my 30s. I have a curvaceous build. I am seeking an honest, sincere man from a farming background for friendship and possibly more. If you are interested please get in touch. Genuine calls only please.

Martin laughed and made a joke of it, throwing the paper onto the sofa. He didn't want his mother to know he'd read the ad earlier in the day and had been thinking about it all afternoon while he was cutting hedges. Did curvaceous mean fat? Or did she just have big tits? Big tits was something he was definitely interested in.

He met a nice girl at the ploughing last year but was too shy to text her afterwards. They got chatting at the Land Rover exhibition. Not that he had the money to buy a new Land Rover, they were only for flashy fellas anyway, not people like him who had bashed-up ten-year-old Pajeros. He got chatting to this woman anyway, and he thought she was fairly down to earth. Then when she was leaving to go to the cookery demonstration with Neven Maguire, she gave Martin her mobile number. He ended up thinking about it too long whether he'd text her or not. Then sure, a week had passed and there was no point; she wouldn't remember him anyway.

He regretted it afterwards. Now he was thinking that maybe he should get in touch with the woman in the *Journal* with the big tits. It couldn't do much harm, and after all, he wouldn't tell anyone, especially not his mam. Just in case nothing came of it. Sometimes he got depressed about not having some kind of female company, but then there was so much to do on the farm and so much to worry about that he didn't give women much thought most of the time. Anyway, there were no women left in his part of the world. They were all off working in Dublin, in college or travelling the world. Half of his parish alone seemed to be in Australia, on Bondi Beach or somewhere like that. And sure, why would they come back, when there's nothing to work at only the shop, a couple of pubs and the garage? Even the nearest big town was quiet now for jobs as well, especially since the technology plant closed down.

On weekend nights the pub he goes to is full of men, just like it used to be after the marts, the bar lined with farmers sitting over pints; them that's left in it. It only hit him that he was beginning to look and sound like an old man himself; one of those fellas with no wife or children, coming into the town once a month to sell a few cattle and then heading back out to their farm with an empty trailer, to one lonely light and only the half-blind sheepdog to welcome them at the door. And that was that. It was only the other day, when he heard his cousin, who must be at secondary school now, say to her friends that he was a 'bachelor farmer', that he thought he needed to make more effort, try to meet someone soon before he turned into someone he used to laugh at as a boy.

How had the years gone by so fast? After Gurteen Martin had come home to farm; he had never really thought of doing anything else. It was good land that had been in the family a long time, his brothers weren't interested, and— as many of the local fellas would have it—a very good beef enterprise fell into his lap. He laughed along and rarely told any of them how hard it actually was. Working day in, day out with his father and living at home looked good to outsiders, but you try doing it for a living. He argued constantly with his father over changes they needed to make if they wanted to keep in farming.

Fellas round the country were now looking into newer breeds such as Piedmontese and Romagnola when it had taken blood, sweat and tears to convince his father to get into Limousins and Belgian Blues ten years ago. It seemed that the minute one thing was making you money, farmers were told to change their systems and do something else. Reinvest, buy new machinery, change the slatted sheds to out wintering pads, go organic, don't go organic. Over the years, the meat plants and supermarkets had squeezed down the profit margins until it was almost not worth it to farm any more. Sometimes at the end of a day at the mart, Martin wondered if it was worth all the effort.

For all the time he put into his livestock, calving, feeding and fretting over them, he got so little in return it would break your heart. But Martin would light another fag, swing himself into his mud-spattered Pajero and head for home. He didn't like to let on how much these things sometimes got to him.

When you looked at TV programmes like *Ear to the Ground*, farming seemed attractive, especially with all these people making home-cured hams and all that expensive stuff that rich people in Dublin and Cork spent a fortune on. Martin's life was far simpler, far more of a struggle, but he still didn't want to give it up. If it's what you're born into it's very hard to turn your back on it. That's what they say.

LUST FOR LAND

Turning your back on a good beef farm is a notion that would wake the dead in many parts of rural Ireland. But in reality, the shrinking numbers of farmers and the flight, particularly of young people, away from the land has been the biggest population trend of Ireland's recent past. Farming is in decline; living off the land has become a minority sport. The unbelievable has happened: land, something the Irish were seemingly obsessed with, has ceased to hold its magic over us.

When John B. Keane's character thundered about the 'law of the land' in the 1930s, he was a product of forces at work in Ireland since Cromwell's time. For centuries, the downtrodden Irish were denied ownership in our own country of the one thing that could improve our lives—land. The consequence of Cromwell's dictum 'to Hell or to Connacht' was that five-sixths of the population had to make do with owning one-sixth of the land, and the boggy, marginal sixth at that.

The unfortunate hangover of land lust in Ireland was a huge economic, social, artistic and sometimes fantastical connection with the land. Somewhere deep in our heads we desperately needed to *have* land, and if we couldn't have it ourselves, we desperately needed that no one else could get their hands on it. And so it came to pass that for generations we accepted that this was the way Ireland was, the way people thought, the subject we wrote, painted and lamented about. Alongside getting rid of the English (in order to get hold of our land), our attachment to land and our countryside origins became our particular way of being; it was the cross we chose to drag around with us while everyone else was suffering pogroms, civil wars or the brutality of French, Belgian or English rule. We had the land issue. Land was the Irish stich. It was stolen from us! We need it, we want it back!

In the sunny days of modern Ireland, which began (as we are told to believe) with *The Late Late Show*, we began to calm down a bit. Our need for land began to wane. All over the country, men like Martin's uncle, Mickey Joe, stepped onto a train bound for Dublin, leaving his townland far behind him to get a job and hopefully begin having sex. Occasionally he thought about saving five shillings a week from his wages to buy O'Dwyer's field down at home, but over time, city life got its claws into Mickey Joe and the field began to lose its fuzzy, dreamlike qualities. It now had to compete with newer, glossier fantasies, such as Mickey Joe as a drummer in a showband and Mickey Joe rolling in the surf with scantily clad American actresses. Yet, in solitary moments, sometimes while riding his bicycle into work, Mickey Joe thought from time to time about O'Dwyer's field. How well drained it was, how much of it was covered in good clover, and how he really didn't want that false, crooked cousin of his from down the road to get his filthy hands anywhere near that field. 'Maybes I should buy it anyway,' he thought. 'Wouldn't it make the oul' mammy all the happier?'

The land was exerting its magic again. Even as we became more urbanised, it still drifted about our brains like the bad fairy on our shoulder, urging us never to forget our attachment, our dependence on it. To outsiders it all looked a little bit weird. From time to time we found ourselves explaining to curious foreign onlookers, Yes, land is a bit of an obsession in Ireland. Yes, a man killed his neighbour in order to get his hands on the field next door.[1] Yes, it's all a bit sad, but at least we're not from Holland and obsessed with weird sex and clogs.

CELEBRITY MAKEOVER

But the shine is beginning to lose its lustre, the genie is out of the bottle and well and truly departed. Land just doesn't excite the same passion any more, and if it does, it's for a very different set of reasons. In Ireland's very recent past, land has gone through a celebrity makeover. No longer content to look dowdy and support a few ragged sheep and third-rate heifers, land went under the plastic surgeon's knife. It is glammed up, sparkly, sporting glass-balconied apartment buildings, hotels with outdoor hot tubs, holiday homes that look like Irish cottages but were designed in Finland. Rural Ireland has gone through its mid-life crisis makeover and has come out the other side sporting a gigantic boob job and a cantilevered bra.

For the past fifteen years, a parcel of land for sale excited an immediate spike in the blood pressure of those who saw it solely for its development potential. And even now, with the property market in tatters, a half acre of scrappy land can still give birth to twenty apartments or twelve semi-ds when the time is right and the planning favourable. Buy it, hold on to it, then develop the crap out of it later on. Not just in urban Ireland, but also in rural areas, our perception of land is one of a cash cow for some kind of transformation; a plot for wooden holiday cabins, a golf course. The value of land today lies in transforming it into something entirely different from what it actually is.

Land as somewhere to produce food is something becoming rapidly more alien. Our remove from the production of food has accelerated this loss of connection. While we may joke that Ireland's obsession with land is something unique to our messed-up, colony-whipped minds, land obsession is far from unique to Ireland. It's something still of huge importance to the 1.6 billion people around the world who still rely on daily toil on the land for their survival. While we may have viewed ourselves as unhealthily land obsessed, the need for land was primarily about the need for food.

Was it the famine that really exacerbated this in Ireland? The shared cultural memory of having no food, of starvation and death? All the more reason to have a healthy respect for how essential land is to our survival. Irish adults today are only three generations away from living memory of the famine. And even if the fear of it is not still floating around somewhere in our DNA, our early school education drummed the story into us pretty well; James Mahoney's drawings peppered our history books with hollow-eyed victims, barely clothed, and children dead at the side of the road.[2] How quickly we have moved on to become a population that never thinks about food security. How would it ever run out? Aren't there supermarkets, for God's sake?

At the time of the famine, the English economist Nassau Senior observed that he expected the famine to kill 'not more than a million people, [but] that would scarcely be enough to do much good'.[3] Hardly a sympathetic observation, but many at the time viewed the famine as a Malthusian instrument which would rid Ireland of its high population, free up food sources and improve living standards. Indeed, in the years following the famine, per capita incomes in Ireland rose more rapidly than those of our English counterparts.[4] Continued emigration and falling birth rates after the famine also helped improve the welfare of those who had survived it. Nevertheless, the lasting impact of the 1840s left a residual fear hanging over subsequent generations; an antidote to the chief villain, *Phytophthora infestans*, was only discovered in the 1880s. Our great-grandparents' parents were adults then, most of them living or farming in rural areas around Ireland. The fear of the same thing happening again can't have been far from their minds. 'The bit of land', if you were lucky enough to own some, enabled families to become food secure, to grow a surplus of food for cash, to create opportunities for their future. Land holdings became more substantial following the famine, and tillage farming declined, as did marriage.

In 1870, when British political opinion at last woke up to the need for land reform in Ireland, only three per cent of the population owned land. By the time Pearse and Co. hoisted the tricolour over the GPO, that figure had jumped to seventy per cent. Now the trend is reversing, with land ownership falling into fewer and fewer hands.

Where did it change? Where did hundreds of years of psychological conditioning disappear to? Conventional wisdom remains that we have a lust for land like no other nation in the Western world. This is plainly no longer the case: land, particularly if it has no development potential, fails to attract our interest. In the last twenty years, the number of farms has shrunk by almost twenty-five per cent. While land may still have an affectionate home somewhere deep inside our brains, we are leaving it behind, in both real and psychological terms.

THE LESSER SPOTTED FARMER
After almost sixty years of dry stock and tillage farming just outside Ballinasloe in Galway, Ignatius Colohan sold up in 1996. 'Turned my back on it, and that was it.' There had been times in the eighties when the price of barley was so high that he thought he'd 'never see a poor day again'. They were offset by harvests that were so bad 'you'd be literally praying all the time'. His was one of the 1,300 farms sold that year.[5] Ultimately, price pressure pushed him out, as year by year what he made from farming got 'thinner and thinner'. Recently he went for a walk with his wife, Pauline, past their old farm. His sheds have been converted into a garage and the land is unrecognisable. 'Have you any regrets about the land?' she inquired. 'Not a bit.' And he sounds like he means it.

Now Ignatius and Pauline are going to add their numbers to the urban migration statistics. They hope to follow their daughters towards Dublin, where they want to build a house. For Pauline, Dublin offers concerts, recitals, shops and the opportunity 'to teach a bit of piano'. Ignatius makes himself useful around his son-in-law's picture-framing shop in the suburbs and there's a tempting array of golf courses he's never played. When he hears fellas talking about prices at the mart it's all a 'foreign language' to him now. His and Pauline's interests have shifted away from the countryside. While they haven't yet made the physical move, they have joined the swelling ranks of urban Ireland, at least in their own minds. Ignatius says that he's tempted to believe some who slag him that he was 'never a real farmer anyway', and it is surprising how a man who lived and worked on a farm since before World War II can make the transition away from rural life so easily.

In Ireland's cities, packed eastern seaboard and growing suburban areas, rural Ireland and the food that is grown there get little attention. While most of us will enjoy a perfectly cooked fillet at a restaurant table, we give little thought to the fact that the very people who produced it, farmers, are a dying breed. Their importance in Irish society is diminishing, along with their numbers; only forty thousand are still farming on a full-time basis in Ireland. By 2025 there may only be a quarter of those still at it.[6] These days it's more usual to find those in farming working part time on their land, with fifty-eight per cent of farmers having another job away from the farm.

In our boom years the buoyant construction sector pulled many farmers away from full-time farming—after all, a bit of plastering or pouring foundations paid better than keeping livestock. Policy developments in agriculture, such as the decoupling of direct payments from production, are also likely to push more Irish farmers from the land. The budget of 2008 didn't help either, with various scheme payments, such as the early retirement scheme, and payments to those in disadvantaged areas cut, creating widespread disbelief in a sector that felt it was on the margins as it was. The Irish Farmers' Association estimated that following the budget, a cattle farmer in a disadvantaged area who had been earning just €21,500 before the budget would now be taking an income cut of eleven per cent. Times were hard, and about to get a bit harder.

As the building sites fell silent and the housing market went into decline, the jobs that farmers traditionally did away from the farm—working in the the construction sector and traditional manufacturing—became precisely the areas of employment most vulnerable in times of economic slowdown. On many Irish farms it is often the farmer's wife who is keeping the ship afloat. Especially in the current climate, the situation for farmers' spouses is more optimistic, as many of them are employed in professional and associate professional jobs in the public sector, such as teaching, nursing and administration. These jobs are considered more 'secure' in the medium term. As farm incomes declined and the building boom collapsed, it's income from the female side of the family that's keeping many Irish farming homes going.

THE ONLY GAY IN THE VILLAGE

So if farming is less popular as an occupation, is there anyone out there who still wants to enter the land of mud, Massey Fergusons and yearly protests outside Leinster House? Apparently so. Macra na Feirme, rural Ireland's answer to Wesley Disco, has eight thousand members in three hundred clubs throughout the country. Approximately one-third of Macra members are involved in farming, with males making up sixty per cent of its membership and females forty per cent. They say there is plenty of optimism for the newer generation coming into farming. A recent survey conducted by Macra of 109 young farmers indicated that sixty-seven per cent of those surveyed were optimistic about the future of farming in Ireland.[7] According to Teagasc, enrolment figures for further education courses in agriculture were up by forty-two per cent in 2008.[8]

Their optimism is encouraging, but the longer-term trend is undeniable. While the Central Statistics Office recorded the existence of 133,000 farms in the State in 2005, this represents a decline of twenty-two per cent on the 1991 figure.[9] Just eight per cent of Irish farmers are under the age of thirty-five, while over one-quarter are entitled to a bus pass. The people and the industry that shaped the way the countryside looks are slowly passing into the night. So who lives there now?

The last thirty years or so have seen huge changes in the type of people that make up rural Ireland; new communities of Brazilians, Polish and other migrant workers have come to parts of Ireland that used to regard German holiday homers as alien. Some areas have seen a huge drain on their younger populations. Other counties have become transformed into commuter belts for Cork, Galway and Dublin, with thirty per cent of people in rural Ireland commuting thirty miles to work.[10] New populations and community structures have created social stresses in some areas, including access to schools, public transport and health services. Rural poverty is still a reality in Ireland, and so is isolation. Twenty-one per cent of rural households are people living alone, and the worry of suicide and isolation among those living alone is high on the agenda.

The populations of rural areas are declining at unequal rates. Suburbanisation of countryside areas has reversed migration in some places, but elsewhere the population has been decimated. Those left behind tend to be socially and economically isolated elderly men.

A map of Ireland in 2006 depicting the number of people per thousand aged over sixty-five reveals that most of them live in the North-West.[11] A surprising forty-two per cent of us in Ireland still live in rural areas where the population is less than a hundred and fifty people per square kilometre. Where population growth has happened in rural areas, it has been driven by inward migration; these areas have a younger demographic than was the case in the recent past, and their age structure is significant for the development of rural areas. Keeping and integrating these migrant rural communities is crucial to maintaining life in some areas of Ireland that will otherwise be almost entirely denuded of people.

As Ignatius and Pauline Colohan prepare to say farewell to Ballinasloe, Ignatius notes that they are only part of a trend. Where he once farmed eighty acres, few in the area are now farming anything smaller than three hundred acres. Do the maths and you can see how this translates into literally two-thirds fewer farmers in the town's hinterland.

And what about our farmer Martin's sex life? It still isn't looking too good. One of the big unchanging factors in rural Ireland is the flight of women towards urban areas and the high number of single males left looking for partners in an increasingly small pool of available females. At least we now have the internet, and in areas where social interaction is becoming more and more reduced, the power of online communities and online communication, dating and interest groups has filled a huge gap.

For gays and lesbians in old Ireland, living in a rural community pretty much put an end to your expectations of ever finding love. While this is still an issue where great prejudice abounds, overall, attitudes may be improving. 'But there are still some dark corners,' says Ciaran Reilly of Gay and Lesbian Action Midlands (GLAM), a group organising social events predominantly in Athlone and Mullingar. 'When people do try to contact us there still is a strong element of fear, of being unwilling to give any information,' he says, pointing to a discrepancy between the development of gay scenes in rural Ireland and in the major cities. People who do attend GLAM social events look for reassurances that the group is mixed and not overtly camp, Mr Reilly says. 'There's a reluctance to go out and meet other gay people on their own doorstep. They'd actually rather go to Dublin.'[12] Other organisations connecting gays in rural areas are Gay Clare and OutWest. All these networks have been hugely helped by internet communication, but for many, coming out in a rural community or a small town outside Dublin is still not an option.

In the straight community, meeting a mate in rural Ireland is getting more difficult. Results of a survey carried out at the 2008 Ploughing Championships (where Martin could have scored) revealed that forty-four per cent of farmers met their spouse at a social event, with most in the under-forty age group saying they met in a pub or through a club such as Macra. In the forty to sixty age category, almost half had met their spouses at a dance, while fifty-eight per cent of people aged over sixty cited the dance hall as where their romance began.[13] As the dance hall disappears as a meeting place, the internet and 'meeting a partner through work' were categories that in previous generations were unknown to young farmers looking for love.

But meeting someone at work reflects the off-farm interests of many rural Irish who want to keep a bit of farming on the go while knowing their income in reality is going to come from somewhere else. Like the Colohans, the advantages they see of living in a city or urbanised area are tending to outweigh their ties to the countryside. For others, the pressure to leave is not social but economic. Farming doesn't pay the bills and unless you invest, diversify or win the Lotto, your projected income bracket over the next ten years could well be in the lowest category in the country.

WHAT'S DOG KENNELLING GOT TO DO WITH PRODUCING FOOD?

As farming itself undergoes huge change, does the sector represent a safe bet, or is it a dangerous rollercoaster ride for those willing to stay in the countryside, invest in the future and put the work in? As incomes have fallen across the board for the traditional types of farming—dry stock, tillage and dairying—farmers have been urged to diversify to survive, perhaps into rural tourism and recreation, or the production of high-value-added food products (artisan cheeses, preserves, etc.) or organic farming. But in research, farmers have at times expressed scepticism, disillusionment and estrangement with the changes they are being forced to make in order to survive in a business that is constantly changing. 'All the investments I've made in the farm in the past are now obsolete. Now we're expected to make more investments, and to change everything. But I'm not going to do that as I'm finished with it, I'm retiring now.' This view of one is typical of many.[14]

Many farmers feel that diversifying into B&BS (a troubled sector as it is), organics, etc. is not an attractive option, and most opt for off-farm employment instead. For many years, EU funding through the Leader projects has been available to rural development schemes, many of which have been started by farmers looking for another option, such as starting a goats' milk enterprise, building a horse livery yard or converting their farm buildings into a rural adventure centre. One Wexford farmer made such a move by adding a dog kennel business to his beef farm. Before opening his boarding kennels, he calculated, he had been making forty cents a day on his best bullocks. He now charges €12 a day to board each dog. It doesn't take much to see how kennelling dogs is a better earner than producing food: it's just a bit of a surprise.

In 2008, €425 million was given out to these kinds of farm diversification and rural development schemes. Éamonn Ó Cuív, Minister for Community, Rural and Gaeltacht Affairs, said the availability of such a sum in Leader funding could leverage a further €600 million of private funds for rural development. 'There are more unexploited resources in this country than we can develop, and despite the odds and what's happening on the world scene, I believe this programme will be a great opportunity for rural Ireland.'[15]

If farmers are now doing something other than farming, it's clear that farming is in a state of flux, or in some cases in decline. Alongside this change in the countryside, there are other changes afoot that threaten the nature of rural Ireland's identity itself. How much of the Irish character is rooted in a rural culture? If this culture undergoes radical change, what happens to the Irish character itself? In the past, farming, the land and the Church were the big shapers of rural society (they were also fairly potent ones even in urban Ireland), but if things are changing in the countryside, will these changes have the same impact on our national psyche as they would have a generation ago? Many Irish people feel that there has been a chipping away at rural Ireland itself in terms of the erosion of services to communities and the lack of incentives to stay there. The closure of post offices and rural pubs have been seen as the biggest losses to rural dwellers. The Vintners' Federation of Ireland estimates that one thousand pubs have shut down over the past three years. These closures are not just about access to alcohol; they are felt in changes to the way a rural community interacts with its members. 'People feel like prisoners in their own homes,' says Dick Dunne, a publican in Stradbally, County Offaly. 'There is a lot of anger in rural Ireland about this. Post offices are also closing. So there are fewer places where people meet.'[16]

When RTÉ closed down its eighty-two-year-old medium wave service, panicked callers, whose radio dials had been glued to RTÉ Radio One by a combination of dust and grease from the Stanley range, rang RTÉ wondering how they were going to tune into the new, frightening-sounding FM service, despite the fact that it had been around for about four decades. Poor medium wave, it's gone the same way as cutting your own turf, the Child of Prague and heading to the pub after Mass on Sunday. How RTÉ expects an increasingly elderly rural population to adapt to digital radio and to cruise stations tailored for their age group is as curious as the mysteries of Fatima. While we know the Ireland of old is shrinking, we are, in many cases, doing very little to prevent further isolation happening or to help mesh together the new and old types of rural experience.

At the 2008 National Ploughing Championships in Kilkenny, the *Irish Farmers' Journal* canvassed rural people about their current concerns. Health issues, particularly cancer misdiagnosis and MRSA, featured at the top of the list, and those questioned also talked about the downturn in the economy, job prospects for children and climate change as real worries.[17] If these are the concerns of rural people, they also reflect quite closely the concerns of Ireland's urban population.

Are the two becoming less distinct? As farming activity declines in rural Ireland and a higher proportion of people have the same types of occupation as urban people, does it mean we are more connected? While so few people in rural Ireland still work in farming, the rest are still busy working away in shops, garages, insurance offices, hospitals and the public service, just like everybody else. As farming declines, are we therefore forced into sharing the same fears, and are there fewer differences between the way the two communities live?

This lack of difference has both good and bad aspects for Ireland as a whole. If more of us share a common set of experiences it should make it easier to lobby for a better way of life across the board in Ireland, rather than relying on what often seem self-interested campaigns for fractured community interests. If we all lived fifty miles from a Boots, an M&S and a Tesco we might act more quickly to save Irish businesses and livelihoods; perhaps this would no longer be (as it was until recently) an east coast debate. Likewise, if rural children, who are increasingly suburbanised, are less physically active and more glued to computer games—just as urban children are—it makes thinking about health policies and safe environments where children can play and exercise something that legislators across the country have to think about.

On the other hand, farming in Ireland will become a shrinking activity which will increasingly be split into two camps. The first group will be the people who farm part time and/or on a small hobby farm basis. They will in some cases produce specialist produce—organic veg, artisan or value-added foodstuffs—or just keep a few cattle around the place because it's what their family always did. The second tier of farmers will be those who have adapted and grown hugely in size in order to stay profitable: the big dairy operators, where having a thousand cows will not be unusual, or the big tillage farmers who not only grow large acreages in Ireland, but also have farms in Eastern Europe, America and the developing world. Farming will be less a way of life as it becomes big business, confined to those who know how to do big business well. Farmers will be well educated, worldly and market savvy, unlikely to be seen hanging over the rail at Kilkenny mart, but running businesses that are connected to the food-producing electronic highway, an environment that is highly dependent on technology and not on whether your neighbour has chosen to spray his potatoes today or not.

The price we will all pay for the shrinking number of farmers and, more importantly, their shrinking importance as part of our culture is an increasing distance from the way food is produced. Even now, the milk from our local dairy farmer is as likely to be exported to Malaysia as to go into the Tetra Pak we buy at the supermarket. The interconnectivity between Irish people, land and food has already taken a battering. As we lose our way of life in old Ireland we are becoming more worldly, but we have less of a relationship with the food that we actually produce here.

In our rush towards the future and a more globalised, smaller world, we must be careful not to throw out the baby with the bathwater. The collapse in the value of property is forcing us to look at land itself with a new eye. Hopefully this will allow a re-engagement with the debate on what the future of rural Ireland should be. For a few years there we lost the run of ourselves. We forgot about growing food, about keeping communities afloat, about maintaining a sustainable environment for our children and the generations following them. We forgot about jobs that were based on actually making or growing something rather than changing land into building sites. In an era where we need new ideas fast, it might be worth turning to some of the older ones for guidance.

There has been a decline in rural Ireland, but there is also still a fair degree of vibrancy. Whether it's the GAA or the willingness of rural communities to help each other in times of crisis, there are lessons to be learned from the good aspects of 'old' Ireland that managed to survive a period where their very existence was deeply unfashionable and, at worst, woefully neglected. We took rural Ireland for granted. We plundered it as a vast larder for food, leisure exploits, motorway building and property speculation. It might now be time to give something back and support the farmers and food producers who are both custodians of the land and providers of what we eat. There are still problems with rural Ireland, but if we turn our backs on it, we might all regret missing the community, environmental, food-producing and huge economic opportunities that rural Ireland has to offer. Who is rural Ireland? To a large degree, it's all of us.

Chapter 5 ❧

| THE URBAN PLAYGROUND

Decadent and doomed

In the opening sequence of the Merchant Ivory film *The Remains of the Day*, the camera sweeps across a ridiculously picturesque panorama of English landscape. Horses and riders are scattered across open parkland; cantering behind a pack of moving hounds, they follow the call of the hunting horn towards the giant honey-coloured facade of Dyrham Park. The tableau is one that director James Ivory has a particular genius for creating, a landscape rich in cultural detail that immediately tells the audience all they need to know about the unfolding story's place in time and society.

In this case, the opening frames of the film paint an instantly recognisable portrait of English country life between the wars. It is an image that is both romantic and decadent. Unfortunately for the horses and riders of the Beaufort Hunt who featured in the opening sequence of the film, it was this very portrait and the meanings attached to it that was to be their downfall. Twelve years later, the Beaufort Hunt became a casualty of a changing Britain. The Labour government brought an end to a countryside activity in which humans had engaged for thousands of years: hunting with dogs.

Hunting was perceived by many in England as a barbaric activity of the upper classes that was long past its sell-by date. The bill on hunting was largely supported in England's towns and cities, but in the countryside it was seen as threatening to strike the final blow in what had become a huge clash between urban and rural culture. Country people felt the proposed ban on hunting exhibited a fundamental misunderstanding of their way of life and a final kick in the face to the shrinking number of people still trying to make their living in Britain's rural economy. In the run-up to the bill's passage, what surprised many in England was the strength of feeling evoked by the plan to ban hunting and the level of fierce opposition to the bill. Emotive television debates, protest marches and physical violence between animal rights activists and rural people were regular events.

Hunting, once described as 'the unspeakable in pursuit of the inedible', had become a politically charged hot potato which many politicians simply hadn't seen coming. The ban was the symbolic tipping point for hundreds of thousands of people in rural England who felt that in the world of New Labour, they were hidden from view or, at worst, their way of life was being slowly wiped out. The enormous countryside marches on London, the largest of which saw over four hundred thousand people protest in the centre of the city, was the crystallisation of a new social and ideological protest movement. 'The March for Liberty and Livelihood' was, in effect, not really about hunting, but about the right of rural people to sustain their way of life. Marchers claimed that their protest was 'a cry for help as well as anger'.[1] They wanted not just government but also the people of Britain's towns and cities to sit up and listen. Most of the people who marched on London did not actually hunt themselves. The factors that led them to protest were a set of circumstances that had been brewing in England over the past decade or more. The ban on hunting was simply what brought it to a head.

WONDERWALL

We may think in Ireland that the change from a rural economy to an urban one came quickly, but England experienced a much more profound alteration in its distribution of population, with change from a rural to an urban nation happening more rapidly. By the early twenty-first century, farming's economic contribution had been falling for decades, with the number of businesses related to agriculture, forestry or fishing falling by fifteen per cent between 1994 and 2004. The National Farmers' Union pointed out that this was due to the fall in the number of medium-sized family farms. These have either been gobbled up by larger farms or broken up to become part-time smallholdings or extended grounds for an adjoining farmhouse. Forty years ago farming made up three per cent of Britain's GDP, but now it accounts for just over one per cent.[2]

England's countryside dwellers had a long list of grievances to bring to the countryside marches on Whitehall. In 2002, the Countryside Agency Report found that rural wages were on average twelve per cent lower than urban ones. Also on the dissatisfaction agenda was the closing of five per cent of rural post offices, house prices that were out of the reach of rural dwellers, the closure of banks and services in remote villages, and the overall conviction that the Labour government wasn't listening to their concerns. Two years before the marches on London, the government's rural White Paper said that all departments would take account of rural needs when making policy. England's Countryside Agency (CA) was asked to monitor this 'rural proofing' but found that eighteen months later, government policymakers were still not 'thinking rural'.[3]

The EU's Common Agricultural Policy left many farmers in Britain feeling that working for subsidies or maintaining land that in some cases they were paid to not farm at all was against their ethos. It was also blamed for falling farm incomes in general across the UK. Perhaps British farmers found the EU's farming regime more difficult to bear because per capita they gained less from the system than their counterparts in Ireland. It was also the case that many had had a better standard of living than Irish farmers in the first place and were thus less eager to embrace the CAP and its implications.

With many farmers being typically Conservative supporters, under the shiny new leadership of Tony Blair they felt economically, politically and socially on the margins. The poster boys for Blair's Cool Britannia were Mancunian pop stars and people who made a living from standing on red carpets in designer clothes. Farmers they were not.

The trench between rural and urban life in Britain was getting wider. It didn't help that in 2001 the food and mouth epidemic decimated farming in the UK and was viewed by many as the final nail in the already polished coffin of British farming. Labour's mishandling of the crisis had helped the epidemic grow to mammoth proportions. The failure to act quickly was seen by many as an indication of their lack of connection with rural life. Many farmers blamed the crisis on the Labour government not understanding the scale of the disease and their subsequent failure to put control and quarantine measures in place quickly enough: the starting point that led to horrific pyres of hundreds of burning cattle.

SUZANNE MEETS SHEEP WITH NO MOJO

In the aftermath of the foot and mouth crisis I visited a farm in Cumbria that had lost its flock of sheep to the disease. Animals which had grazed hills at the margins of the Lake District for hundreds of years were now absent. With them disappeared a genetic heritage and what Welsh hill farmers term 'cynefin': a flock's sense of place or territory that is passed from one generation to another, a kind of instinctive knowledge that is key to a herd's continued survival in its own particular environment. This family's herd had been rounded up and burned after the infection had spread to their area from Carlisle mart. After the crisis, many farmers in the area were choosing not to restock, having lost irreplaceable bloodlines and genetic heritage. Some were simply broken-hearted, unable to begin rearing more livestock after the losses they had witnessed.

After spending a day with this community I came away with a feeling of intense sadness. There was a palpable sense of despair about their households, their daily routine and way of life now utterly changed. One family sat at a kitchen table looking out on fields and pens that were gapingly empty. To me they seemed like people without purpose, who had lost their way in life. We had been so much more fortunate in Ireland; our animal culls had been a fraction of what English farmers had experienced. Did our government move faster because they were closer to rural culture, or was Ireland's experience of foot and mouth just a lucky accident?

Following the crisis, the UK's Ministry of Agriculture, Fisheries and Food was renamed DEFRA, the Department for Environment, Food and Rural Affairs ('farming' doesn't get a mention). Even the name change was regarded by British farm leaders as another kick in the teeth to farmers; they felt that Labour deemed a change in name was enough for people to forget what had happened.

In the background, the bill on hunting was rumbling ahead, and as a hunting ban began to look like a real possibility, feelings were brought to a head once more. Labour supporters and many urban Britons were gearing up for an end to a barbaric sport engaged in by rich people with an ill-placed bloodlust. In preparation, the Countryside Alliance gathered its opposition troops and mounted a huge defensive campaign, saying that hunting was an effective method of pest control that supported a rural economy ranging from farriers to vets, hunt staff, livery stables, feed merchants and so on. Take away hunting and you take away our livelihoods, they said. Take away hunting and you destroy our right to keep alive customs, skills and countryside knowledge that have been practised for generations.

The marchers hit London and the debate reached fever pitch, but ultimately the protest failed. The Hunting Act was finally passed in 2004. In the heightened emotive atmosphere, award-winning documentary-maker Molly Dineen made a film entitled *Lie of the Land*, which followed the lives of farmers struggling to come to terms with the ban on hunting and to farm in an increasingly difficult climate. The farmers felt that the Britain they lived in was nearing the end of producing its own food; farming itself seemed to be turning into an expensive minority sport, a casualty to a world trade in cheap and instant food. Britain's reliance on cheap imports and the loss of standards in food produce consumed by disconnected urbanites seemed to them to herald the beginning of the end. The film quietly served up the raw emotions and alienation suffered by many rural people in that period; Britain's agricultural way of life was firmly on the decline, and they were slowly sinking with it. It was a powerful film and one that delineated the strength of emotion and feelings of anger and abandonment that many rural people were experiencing at the time.

Six years on from the ban, the hunts are still getting used to the new regime. Most continued under the new regulations, which are in themselves complicated, have landed many hunt staff in court and are difficult to police. The negative financial impact on Britain's equine and hunting sector is said to be less than anticipated, but the ban has not stopped tension between animal rights activists and hunts or healed the urban and rural divide. Some people in Britain still want to see the sport stopped for good. Could it happen here? Is the gap between rural and urban as pronounced in Ireland as it is in Britain? Is it the case that in Ireland we are heading in the same direction, but at a slower pace? Or do we still have something uniquely rural in our subconscious that is preventing the same sense of separation, lack of understanding and anger?

I HAVE A 4x4 AND I'M GOING TO USE IT

While hunting was the issue that brought the rural/urban crisis to a head in Britain, what is certain about Ireland is that there is already a power struggle between various factions who have influence over how the countryside is run and who want to control it. In terms of wider issues, such as legislation decided at EU or government level, many farmers and rural dwellers already feel that the establishment of Special Areas of Conservation (SACS), controls on nitrates, clampdowns on one-off rural housing and continuous wrangling between heritage organisations such as An Taisce are putting the interests of those who don't live in rural Ireland ahead of those who do.

One typical Irish example is the squabble surrounding the threatened habitat of the red grouse, whose breeding grounds on the uplands of Connemara are unfortunately also shared by hill sheep. When the EU introduced a de-stocking regime in 2008 to reduce the numbers of sheep and thus save the red grouse habitat, the plan went down spectacularly badly with the hill sheep farmers. They pointed out that it is they, not the red grouse, that is in need of protection; only five farmers under the age of thirty-five are still farming in the whole of Connemara. Reducing the numbers of their flocks allowed to graze on the uplands would turn an already shaky business into an unprofitable one, and the younger farmers still keeping this type of farming alive won't choose to do so in the future.

Alongside restrictions on cutting turf from bogs, many farmers, particularly in the West, feel that legislation concerning their area no longer has concern for them. Which is more important, they ask: an environment or the people actually living in that environment? It's a dilemma that is increasingly driving another wedge between rural and urban. Following the No vote to the Lisbon Treaty in 2008, polls carried out by the *Irish Farmers' Journal* revealed that disaffection with the handling of environmental issues with regard to farmers and rural dwellers had been a key factor in encouraging many to vote No. In Mayo, the Rossport campaign against Shell and numerous differences over SAC and heritage issues contributed directly to a strong No vote. Mayo farmers in particular feel like a threatened bunch. In hindsight, perhaps throwing two fingers up at the Lisbon Treaty and cherishing your sheep more dearly than a red grouse seemed pretty understandable.

And what about Irish rural pursuits and the control of the Irish countryside as an *amenity?* For many farmers, the very word itself strikes terror into their hearts; the idea of private land having *amenity value* is fairly unpalatable, to say the least. Also shocking is the notion that someone living somewhere else can control the activities, sporting or otherwise, that take place on your own private land. In terms of Ireland's picture, it's clear that there are two distinct camps into which rural pursuits fall. In one camp you've traditional field sport activities such as hunting, coursing, fishing and shooting, which draw the ire of animal rights activists as bloodthirsty slaughter. In the other camp are the growing number of newer sports such as hillwalking, climbing, mountain biking, white-water canoeing, 4x4 off-roading, etc., all of which take place in rural Ireland but are largely the domain of urban people. Of all these sports, it's specifically hunting and hillwalking that have already turned into emotive issues that have spawned protests, and in some cases violence, and calls for more legislative control.

WHOSE COUNTRYSIDE IS IT ANYWAY?

In the last five years, health promotion agencies, TV doctors and diet gurus have been telling us morning, noon and night to get out there and walk more. And so walk we do, but the problem is, where? Most urban or suburban people enjoy a ramble in the countryside, whether it's a drive forty minutes out of Dublin or Galway to a beauty spot for a short dog walk or a picnic, or a more organised challenge such as completing sections of the Wicklow Way, taking the Five Peaks Challenge or climbing Carrauntoohil. The trouble is, as hillwalking gains in popularity, so does the need for actual hills. Apart from lands owned by Coillte and designated forest walks, tracks over private land, many of which are in scenic areas, have become disputed territory between farmers and walkers. Outraged landowners appearing with shotguns is not an ideal afternoon's experience when you're on a country ramble, and the idea of land being 'private' or having access prohibited is unpalatable to many who feel that Ireland's countryside should be open to more than those who have the good fortune to be in possession of land. Farmers have voiced many concerns in opposition to groups such as An Taisce or Keep Ireland Open that campaign for more access to the countryside. Again, it comes back to who ultimately has control over land in rural Ireland, be it in the hands of private ownership or not.

Rural development schemes have begun to recognise this mounting pressure between both sides, and in 2008 payments were given to three hundred farmers who established walking trails under the New Walks Scheme. These areas include the Sheep's Head Way in West Cork, the Blue Stack Way in Donegal, the Suck Valley Way in Roscommon and Galway, and the Eamonn a Chnoic Loop in Tipperary. Under the scheme, farmers are paid for the establishment and maintenance of walks with payments up to a maximum of €2,900 per annum under a five-year agreement. It's a small beginning, but the beginning of exactly what? Continual pressure on farmers to open up their land to strangers, to run the risks of legal action over accidents or of interference with stock, or worse still, a chipping away at the right to control your own land and keep strangers off it? And it's not only walkers straying onto your land that sends farmers' blood pressure through the sky. It's the other really loud stuff that gets on people's goats, the off-roading by 4×4s, quads and motorbike scramblers. Recently, the National Parks and Wildlife Service has started to get heavy, using European legislation in a pilot project in the Slieve Blooms to bring successful prosecutions against suv drivers and others damaging the fragile blanket bog habitat in the area.

For the hunting community in Ireland what happened in Britain was a real frightener. Could it happen here? So far, only the Green Party and the late

Tony Gregory (perhaps Ireland's most urban TD) have proposed opposition to hunting in Ireland. Many feel that it would be such a hugely unpopular and risky political decision that the issue will not be revisited in future years. However, opposition to the sport by groups such as the Irish Council Against Bloodsports (ICABS) and various other interests refuses to go away quietly. Following the creation of the Countryside Alliance in the UK, which organised opposition to the anti-hunting bill, the IFA has followed suit with the birth of IFA Countryside. While the IFA says it is a membership organisation for those rural dwellers who may not necessarily be farmers (and therefore cannot be members of the IFA itself), it's hard not to see the similarities between Britain's Countryside Alliance and IFA Countryside.

Could the IFA be arming itself, and mobilising the rural constituency, to campaign against future Dublin-centric policy? With the exception of Brian Cowen, almost all Taoisigh and senior cabinet members of recent memory have been townies. Dublin and the East are always referred to as the engines of economic growth. It's not hard when you live west of the Shannon to feel that the wind is never at your back and always blowing in your face.

HOLES IN THEIR BOOTS

What may seem like an outdated minority sport has a surprising number of participants in Ireland; at present, almost three thousand people hunt weekly.[4] There are forty-two packs of hounds, which almost exclusively hunt foxes. There are also foot followers and beagle packs. The Irish Masters of Foxhounds Association says that its membership is drawn from a wide cross-section of the rural community, with the greatest number being farmers. Each hunt meets two, three or even four days a week between 1 November and St Patrick's Day, depending on the hunting country available and the stamina of its membership. There are also a significant number of beagle and harrier packs, who hunt on foot; currently, 129 packs of hounds are registered. Most people think of hunting as the preserve of mounted followers, but in Ireland, far and away the greatest numbers of hunting enthusiasts follow on foot and do so with great passion and commitment to the sport. Foot harrier packs are particularly popular in the most southerly counties, with more than forty packs registered in County Cork alone. Many hunt clubs are based in cities and towns and consequently restrict their hunting days to the weekends.

Irish huntspeople say that it's a real community activity, with members of hunts coming from more varied and less privileged backgrounds than is at least perceived to be the case in the UK. And this is borne out in most meets of Irish packs, where mounted followers include farmers, local business and tradespeople and a selection of urban-based or tourist riders on livery horses. There are also those who work in the horse industry, perhaps bringing on point-to-point horses, elderly riders of the old school and the very young on tiny ponies. Like most Irish sports, it's primarily a social gathering which involves plenty of slagging, gossiping and a few people suffering murderous hangovers. After a chase lasting a couple of hours, a few people will have fallen off, sworn heavily at their mounts or fellow riders, and a fox or hare will have been killed. It's more usually the case that the quarry lives to fight another day and the talk on the hack home is of who chickened out of jumping the treacherous drain which, by the end of the day, has become the width of the Shannon.

For Irish huntspeople, telling them that it's a sport for the wealthy alone will have them doubled up with laughter and showing you the holes in their boots. Again, from an urban perspective, the image of hunting as a pastime of the elite still dominates, and while the fashion industry churns out magazine spreads every autumn featuring jodhpur-clad models with hunting crops in hand, it promotes the image of hunting as remote, aspirational and more *Brideshead Revisited* than Borris-in-Ossory.

THE 'C' WORD

In the period leading up to the hunting ban in the UK, the word on everybody's lips, but one they were very much afraid to use, was Class. Everyone pretended that it wasn't an issue; the opposition claimed that hunting should be banned as an animal welfare issue rather than to punish those who hunted. However, in the aftermath of the UK ban, a Labour MP, Peter Bradley, made a startling admission: 'Now that hunting has been banned, we ought at last to own up to it: the struggle over the Bill was not just about animal welfare and personal freedom, it was class war.'[5]

Disappointing (though not altogether surprising) as it may be that some in New Labour would try to cast themselves in the role of class warrior rather than dull old legislator, the real issue is: could it happen here? Do Irish people living in the city and suburbs of our larger cities still see field sports as the preserve of rich, useless Hooray Henrys and Henriettas? There is heat in the argument; two out of every three people questioned in a 2007 survey were of the view that fox hunting should be banned.[6] Such a popular opinion usually finds some politician running towards it, but oddly, this is a constituency in search of representation. Those against hunting don't have a political party on which to pin their views. The fact that Fine Gael and Fianna Fáil have a foot in both urban and rural camps has meant that any anti-hunting sentiment within their own ranks has been hushed up. 'What's that, then? A private member's bill to ban fox hunting, you say? Now be a good lad and put that down over there. Yes, right beside the Greens' plan to cut the speed limit by ten kph and Gay Mitchell's Olympics bid for Dublin.'

Even the Greens have been happy to equivocate about hunting, even when party leader John Gormley got the environment portfolio and was in a position to start introducing bans. 'In Britain the agricultural vote is no more than one per cent of the electorate,' says David Wilkinson of IFA Countryside. 'Whereas here, although it is dwindling, it still stands at about ten per cent. The views of the rural TD are always heard at the top table.'

Field sports organisations point out that their membership crosses the divide. Plenty of city dwellers have fishing licences or are members of a gun club. But on the other hand, the antis are almost exclusively urban based.

If Ireland is splitting into two very distinct populations, one urban and one rural, it will create further tension over who owns and manages the countryside and what takes place there. Farmers want to get on with farming, but increasingly, urban interests feel they should have some influence over how farming itself takes place and how the countryside is managed. The establishment of IFA Countryside is evidence that rural Ireland is already gearing itself up for trouble in the years ahead. However, one thing that might make an attempt to ban hunting an unpopular campaign is precisely the belief, whether true or not, that rural and urban people in Ireland have a higher level of connectivity than is the case in the UK. The issue might prove to be such a pain in the neck that it wouldn't be worth the trouble. Rural opposition to Lisbon and the World Trade Organization are still bringing farmers onto the streets to protest, and the thought of adding a ban on hunting into an already boiling pot is most likely unpalatable to most politicians, even the Greens.

Another factor that may keep hunting off the banned list is the value of the horse industry in Ireland. Although most of the industry has become increasingly less connected to hunting, you'll still find that most Irish trainers, jockeys, showjumpers and horse producers hunted at one stage of their career and see the value of hunting both as a rural activity and as one that is important to the horse sector. The sport and leisure horse industry alone, which is the non-racing element of equestrian activity (showjumping, eventing, riding clubs, riding schools, trekking centres, etc.), contributes in excess of €200 million to the Irish economy and employs twelve thousand people. Equestrian holidays are an important part of the tourist industry, with up to one hundred thousand tourists spending approximately €63.5 million annually on equestrian pursuits. Many tourists come to Ireland specifically to ride with legendary hunts such as the Galway Blazers.

In terms of racing, the figures are even more substantial, with the industry supporting more than sixteen thousand full-time jobs dispersed throughout the regions. Giant breeding operations such as Coolmore contribute to a total figure of nearly €200 million worth of horseflesh exported from Ireland annually, and in 2008 the racing and breeding sectors produced a gross economic contribution exceeding €300 million. Racing is also a big tourist attraction, with eighty thousand tourists visiting Ireland to attend meetings throughout the country. But the downturn has hit the industry hard, with prices falling dramatically in sales rings as the recession took hold. Welfare organisations such as the Irish Horse Welfare Trust began to point out a rise in the number of horses being abandoned or left in fields with little or no forage. The boom times for horse ownership in Ireland had certainly hit a rough patch. If people were having problems paying mortgages and feeding their families, horses were certainly now very far down the list.

Horse Racing Ireland argues that more than ever before, government support and a direct link to funding via the betting tax is essential to support the industry. It's clearly a big employer, particularly in some rural areas where there is little other economic activity. Supporting racing via the taxpayer does not go down well with many urban observers, however. Again, the perception of equestrian activities is that it is the preserve of the wealthy, and frankly, who cares if a couple of breeders, trainers or even racecourses go to the wall? The downturn also came hot on the heels of the disbanding of the tax exemption on stallion profits when a new regime came into force in 2008. For the first time since Charles Haughey was finance minister, Ireland's bloodstock industry was forced to pay tax on its earnings. Again, the support of thoroughbred breeders through tax exemption had been an extremely unpopular measure with many urban voters. Most rural people see it differently; yes, there are the very rich at one end of the sport, but there are hundreds of small breeders with a mare or two out the back of the house who are integral to the industry's survival and to the production of good horses, particularly future national hunt champions.

How Ireland's countryside develops in the future and how people make their lives there will hinge to a large degree on whether urban and rural

Ireland talk to each other. Many country people are sick of outsiders telling them what to do. The countryside, they say, is not a playground or heritage resort for urban people; real stuff goes on there, stuff that they make their living from, and if urbanites can't deal with this they should take a flying leap. It's fair enough that townies who can't see further than their Range Rover Sport and who neither know nor want to know anything about rural matters shouldn't be the ones to decide what goes on there. Ultimately it has to be a conversation between urban and rural people, but as increasing urbanisation and global forces take hold in Ireland, are our old-fashioned links to a rural life shrinking? Will we end a uniqueness Ireland had in its rural/urban closeness and dissolve a marriage of two elements that never should have separated in the first place?

Chapter 6 ⌒

THE LAND OF HEART'S DESIRE

The death of the culchie on telly

A group of mean-looking cowboys sit in a tense circle playing poker. The camera pans slowly from player to player; narrowed eyes glance secretively over tightly held hands of cards. Suddenly one cowboy makes a move, slapping three aces onto the table. 'Triple A Golden Maverick,' booms the voiceover. It seems our cowboys were playing for a sack of milk replacer.

Farming advertisements on Irish television and radio used to be part of daily life. Tackling such Harry Potter-type horrors as warbles, immature fluke and inhibited ostralgia, farming ads offered solutions to the infinite range of pests and diseases that were once part and parcel of our shared cultural experience. Twenty years ago it was commonplace to see farming products occupying the highest-rated ad spots on RTÉ television and radio: the slots immediately before Gaybo pranced across the stage of *The Late Late Show* on a Friday and those preceding Don Cockburn's comforting but ramrod-straight delivery of the nine o'clock news.

Alongside Maurice Pratt and *Dear Frankie*, farming ads are now confined to the archives of nostalgia. If media and advertising reflect the way we live or, more accurately, the way we would like to live, it's easy to see how in the last thirty years the picture we have painted of ourselves has changed. Up until the 1980s, much of our iconography was of a green agrarian country; after all, Ireland was identified as the Emerald Isle. Kerrygold advertisements featured sunlit farmsteads and bowls of buttered spuds, and the talk over the dinner table was of 'takin' the harse to France'. The rural nature of our country was a huge selling point, whether it was postcards of flame-haired children placing turf into baskets on a cutesy donkey or the marketing of our food products overseas, culchie Ireland and its idyllic rural experience was what we liked to sell, both outside and within Ireland.

PORSCHE CAYENNE? YOU'RE WORTH IT

Recent years have seen this concentration of rural images falling away. Why shouldn't it, some might say, aren't we an urban culture now? A grown-up nation of Porsche Cayenne drivers and designer outlet shoppers ready to compete with the rest of the world. Town or country dwelling is immaterial when most of what you're told to do are urban experiences: texting on your mobile, buying KFC burgers, broadband or anti-ageing creams that will change your life and make you a happy, healthier, more successful person. There is little place in our arts or media world for the conversations and camaraderie of selling cattle at a mart. Or worming with Ivermectin. For urban people, these types of experience are so out of the ordinary that they are as unfamiliar as mating Martians.

How we see ourselves as a nation is always evolving, but the past few decades in particular have seen the self-image of both Ireland and its people turn inside out. In previous centuries things were simple; being Irish (for most of the population) was usually a matter of not being English, and trying to get free of the iron grip of the Empire. Land itself was central to this way of thinking: the right to own land, provide for your family and plough your own furrow in life were at the heart of much Irish writing, expression and song for hundreds of years.

However, alongside the political, there were other kinds of expression: love poetry in the Irish language, religious writing, the timeless satires of Swift and writing that emerged from the Big House tradition, such as the novels of Maria Edgeworth. Over time, several different types of experience became part of Ireland's cultural expression, but by the end of the twentieth century, our view of ourselves as essentially a rural people, passengers on the great ship *Emerald Isle*, began to go through radical change.

WHISPERING SMUT IN JOHN WAYNE'S EAR

In the final scene of *The Quiet Man*, the director John Huston (a man fond of a jar and not to be trifled with) ordered Maureen O'Hara to whisper something dirty in John Wayne's ear for the last shot of the film. Huston wanted the American actor to turn and look wide-eyed with surprise at the crazy Irish *cailín* he had finally hooked. She duly obliged, and only one take was needed. Maureen O'Hara still refuses to detail what she said to Wayne but admits that it was of a sexual nature, swearing, or both. Good for you, Maureen.[1]

There are many myths about the jiggery pokery that went on during the filming of *The Quiet Man*: that Maureen and Wayne argued constantly on set, that Maureen broke her hand slapping him in the face (true—she fractured a bone in her wrist when Wayne put up his hand to block hers). It was a lively production, to say the least, and the dynamic between the two lead actors, the ringmaster John Huston and the luscious backdrop of Technicolor Ireland resulted in one of the most famous movies about this country ever made. Released in 1952, the film captured the height of Ireland's image as whimsical and wildly beautiful. If we can imagine the twentieth century as a journey of images that we made of ourselves, *The Quiet Man* is at the apex of where the image of Ireland and being Irish is at its most romantic but also its most ridiculous. For it was at this very point that the image was most at odds with what was actually happening in real Ireland. The 1950s was the period of greatest emigration from Ireland since the famine. According to the Commission on Emigration, which reported in 1954, the main causes were economic, although social, psychological and political factors were also driving rural men and women to Britain and the United States.[2] In the 1940s and 1950s, rural women in particular felt there were better work and marital opportunities abroad. Their departure left behind many villages with large numbers of single men and falling birth rates, just at the time when de Valera, ever one for a bit of whimsy himself, was praising Ireland's 'cosy homesteads'.

Yet to Hollywood, Ireland then was at its most glamorous. Some writers have pointed out John Huston's own awareness that his film *The Quiet Man* was a complete fantasy. Much has been written about a piece of dialogue that occurs when John Wayne's character is picked up from the train and driven by horse jarvey to look at the country cottage of his ancestors. En route, he pauses on a pretty stone bridge, and as he surveys the green beauty of the surrounding countryside he mutters, 'Is this real?' No, it wasn't real. Not for Huston, not for Wayne, nor for anybody living in Connemara at that time. It was simply another idyllic image of rural Ireland starring on the world stage that was far more attractive and palatable than the reality.

STEPPING INTO PIG POO

Fantasy rural Ireland had been alive and well long before John Huston turned the volume up to eleven. Our landscape and its inhabitants had been the unwitting victims of this sort of carry-on for decades. In efforts to secure freedom from the dark forces of the Empire, the land of Ireland and its people were repeatedly trotted out as a shining example of why we desperately needed to be self-governing. Ireland was often female in this picture: Hibernia, the beautiful victim raped, pillaged and stolen by the black-hearted English. In turn, London's *Punch* liked to portray the Paddies as ape-like, uncouth and generally fit for little more than the pigsty. In the cartoon-like battle over who we actually were, titanic male figures such as Cúchulainn and Fionn Mac Cumhaill were also employed to represent the glory of a testosterone-filled Ireland. Hey! we said—we have soft romantic Ireland, but we also had superheroes, spirits and fairies, characters who reached their zenith of popularity in the Gaelic Revival movement of the early 1900s.

If we think *Lord of the Dance* was slightly over the top, it's worth examining some of the plays and theatrical pieces (what would now be called 'happenings') put on by Yeats, Lady Gregory and Co. At the time, this was highly praised stuff. Alongside penning some twenty-six plays, Yeats set Ireland on its Nobel and Grammy award-winning haul by earning Ireland's first Nobel Prize for Literature. The Nobel committee said that his work had given 'expression to the whole spirit of a nation'. But what the committee of Swedes was actually saying was that the spirit of our nation was that of the fanciful and whimsical pastoral dramas Yeats had written for the Abbey, arrant nonsense like the farmer's wife who gets swept up and taken away by the fairies in *The Land of Heart's Desire*. Step in please, Michael Flatley.

As mad as this looks to us now, the view of the country that Yeats, Lady Gregory and the other pillars of the Gaelic Revival peddled abroad and reflected back to us was highly agreeable at the time, at least far more agreeable than when Synge's *Playboy of the Western World* tried to point us in the direction of an impoverished and suspicious peasantry. Blood was spilled by Catholics rioting in protest at this alien Protestant's jaundiced view of Ireland's supposed rural idyll. Sean O'Casey's plays of the 1920s (*The Shadow of a Gunman, Juno and the Paycock, The Plough and the Stars*) offered a more realistic take on the glorification of the nationalistic struggle. He eventually fell out with Yeats over the Abbey's refusal to stage *The Silver Tassie*, the ensuing catfight prompting O'Casey to leave Ireland for good.

In order to provide material for the fairies and ghosts at the Abbey, some of the Gaelic Revivalists decided it might be nice to document what people were actually doing and saying at the crossroads and in the cabins of rural Ireland. Lady Gregory of Coole Park in Galway was one of the first to step away from the granite of the country house and to get down and dirty in the pig poo of the local community. Married to a former Governor of Ceylon, who was thirty-five years her senior, Augusta Gregory was a well-travelled lady of liberal outlook who had campaigned for Egyptian nationalism before her attentions turned fully to Ireland. Following a trip to Inis Oírr in 1893, she developed an interest in the Irish language and began to collect stories from the villages close to Lough Conn. She recorded tales and folk material from local people, including those from inmates of the far from glamorous workhouse in Gort.[3] Her interest in Ireland's history, as told by the people of the West, was to lead her to becoming a big supporter of Irish nationalism. Later, her fondness for Irish folklore, employed in the nationalist cause, was to be parodied by both Joyce and Flann O'Brien.

Whether or not she fell too deeply in love with the fairies for her own good, it could be argued that Gregory was one of the important figures in the early collection and documentation of Irish folklore. Scores of folklorists later undertook similar but more valuable work, but Gregory is the one we tend to remember. Considering her background and her social and political milieu, she was a woman far ahead of her time. What she probably didn't know then was that the version of Ireland created by her, Yeats and others hung around like a bad smell long after they themselves had gone to the land of the fairies. While having a large role to play in giving us both Croke Park and TG4, their Gaelic Revival movement was overly romantic and quixotic, rooted in the past and giving little direction for how Ireland was to shape itself culturally in the twentieth century.

NO, FATHER, WE WEREN'T ROLLING IN THE HAY

After 1922, what happened to our picture of ourselves? Was it a more confident portrait once we had attained independence, losing those pesky six counties along the way? Was our image of ourselves in the latter half of the twentieth century that of a more confident people? After all, now we were free to paint, write plays and novels about things other than nationalism. And we did. Joyce and Beckett became the champions of modernist literature. They were world figures, and while they wrote about Ireland, Ireland itself was not central to the work.

Much of the twentieth century saw our self-image still rooted in the story of ourselves as a rural people, not an urban one. But as time rolled on, this expression progressively became a more negative one. Life in rural Ireland post-independence could no longer be viewed through rose-tinted glasses with a greenish hue. The message increasingly moved towards a more realistic but uncomfortable portrayal: that real life down on the farm was pretty hard to swallow.

When Patrick Kavanagh came to Dublin in the 1940s, he was viewed as a very roughly hewn outsider by those in polite Dublin society. Just for good measure, he wasn't too fond of them either. His work received a less than enthusiastic reception; *The Great Hunger* was seized by police on the orders of the Minister for Justice for its attack on the oppression of rural Ireland by the Catholic Church. *Tarry Flynn*, his classic and uncomfortably true account of life on a Monaghan farm, documented the experiences of a young man desperate to escape from under the thumb of parents, the Church and an unyielding land that seemed to drown him in despair. It's a book that marries tremendous comedy with enormous frustration, and still rings bells for anyone who was ever a teenager or who didn't get the girl or the boy. But quelle surprise, it was also banned by the authorities. Yes, we could write about ourselves, but say anything negative and you were bound for hell in a handcart.

By 1966, when *Tarry Flynn* had made it off the black list and onto the stage of the Abbey, voices like John B. Keane's were also coming forth with uncomfortable social truths. *The Field*, with an altogether darker, more ambiguous ending than its big screen adaptation, was first presented in 1965. Even fifteen years later, Keane's depiction of a lecherous bachelor farmer, made celibate by force of circumstance, in *The Chastitute* served up uncomfortable home truths for many on opening night in Cork Opera House. While we wanted tales of countryside Ireland, these stories still had to be sanitised. Perhaps Ireland at this point wasn't mature enough to cope with the warts-and-all experience of our own lives up there on stage or screen, even if we knew they were like that in reality.

In the sixties, while the rest of the Western world was all hanging out with the Beatles or off having sex with Mick Jagger (where did he get the time?), most Irish teenagers were still busy bringing in the hay or attending missions at the local church, where black plastic babies with slots in their heads were shaken up and down as a reminder of how lucky you were to be born under the protection of the great white Catholic Church. The sixties also signalled the arrival of television, and rural life suddenly came into the room via the rickety black and white set in the corner. In the era of one-channel TV, the opportunity to view ourselves on television was a powerful one, and many of RTÉ's early offerings, in that more mono-cultural climate, gripped the national consciousness.

Many Irish people over forty will remember, like the death of President Kennedy, the earth-shattering moment when Maggie and Benjy decided to use contraception on the RTÉ's television drama *The Riordans*. Most good Irish mothers would have jumped across the room and unplugged the TV set before the unutterable word was even heard, but in many homes the cat was now out of the bag, and *The Riordans* raised countless issues that were far beyond acceptable discussion in late 1960s Ireland.

The Riordans first aired in 1965, following the success of RTÉ's urban drama *Tolka Row*. Realising the essentially rural nature of Ireland and the huge popularity of radio programmes with rural appeal, *The Riordans* was ground-breaking on several levels. The series, revolving around a mythical farming family in County Kilkenny, was shot on location by RTÉ's outside broadcast units in Dunboyne in County Meath. This was a technically challenging and expensive production technique at the time and offered a far more realistic rendition of rural life than what could be constructed on a sound stage.

So successful was the location-based production model that it was studied by Yorkshire Television, who went on to launch their own rural soap in the early 1970s— *Emmerdale Farm*. The fact that *Emmerdale* has survived the cull of rural offerings on UK channels is a worthy tale of survival: the combination of outrageous disaster devices (an aeroplane crashed into the village in 1993) and completely unrealistic but enough-to-keep-you-interested plot lines has kept the show still rating highly for ITV.

The Riordans wasn't afraid to tackle the big-hitting issues of the time: sex, marital break-up, poverty, the Church and the pill (cue Darth Vader breathing). Looking back from our time and place at these subject matters, it's hard to believe that most were simply not up for open discussion, let alone aired as dramatic scenes played out in front of your assembled family. The pill storyline (which couldn't have been more innocent, considering that Benjy and Maggie were married and that a second pregnancy would risk Maggie's health) caused huge controversy and brought the expected appalled response from the Catholic Church. The series began broadcasting at the same time as the coming into force of the Succession Act, which for the first time allowed a wife to inherit a farm automatically if her husband died. Before the Act, widows and their children could be left with no means, as a farm could be willed outside the family to a stranger. Until the 1970s, Irish wives were also in the happy position of being unable to open a bank account without their husband's approval. In such a climate, it's no wonder that figures such as Maggie Riordan were as shocking to some as Pamela Anderson naked on her bedroom swing.

In its fourteen years on air, *The Riordans* underwent many changes, latterly resulting in Benjy disappearing mysteriously to work on the missions in Africa (actor Tom Hickey had decided to leave the series) and the arrival of a smouldering Gabriel Byrne to chew hay in the corner of his mouth with much sexual menace. RTÉ axed the series in 1979 and in its wake came *Bracken*, also starring Gabriel Byrne, and then *Glenroe*, in which rural was already becoming more urban, and town life itself more of a player.

The mythical village of Glenroe was situated on the fringes of Dublin and the location filming was carried out in Kilcoole, County Wicklow, a town which itself was transformed in the pattern that *Glenroe* was experiencing. There was far less farming going on locally, and characters now lived in housing estates. A farm shop in the town sold Dinny and Miley's produce, and the new urban world was beginning to dominate the older rural one. *Glenroe* met its end after eighteen years in 2001 and, like the demise of *The Riordans*, the decision to axe the programme angered many. Rural viewers felt that there was decreasing representation of their lives on Irish television. *On Home Ground*, a drama series about a GAA club, ran for two seasons in *Glenroe's* Sunday night slot, but failed to capture the popularity of its predecessors.

Alongside RTÉ, Irish filmmakers were making us aware of more uncomfortable rural experiences in films such as Bob Quinn's powerful *Poitín*. Throughout the seventies and eighties, rural Ireland, the Gaeltacht and farming life were still subjects of relevance to filmmakers and writers. But after the demise of *Glenroe*, what is the experience of rural Irish life that we currently offer up about and to ourselves? Is it the fictional midlands town of *Eden*, where an outwardly happy family suffer lies, personal grief and separation? Is it the ribald comedy of *Killinaskully*, where the remains of Father Ted's surrealism combined with comedy cliché are alive and kicking?

In 2007, Louis Abrahamson's award-winning film *Garage* came very close to putting its finger on what life in a small rural town currently looks like. The story of a garage attendant who innocently walks himself into tragedy is quietly gripping, and the central character brilliantly drawn, subtly directed and sympathetically played by Pat Shortt. However, life in the town outside the focus of the central character, the world around him, is clearly a changing rural Ireland of which the viewer is allowed small glimpses. It's a place where there's 'a fierce amount of buildin' happenin' up in the town and new houses poppin' up every day around the lake'. Newcomers are moving in and business is good—for some. Rows of men sit along the bar, communicating in banter that is both affectionate and violent. Things are changing but remaining the same. Like the stage play *Eden*, the idea of the rural male is questioned and explored. There is sexual tension in the air and the unspoken conversations are those revolving around power, money and employment, or the lack of it. This is a rural town with farming on its fringes and little economic activity, other than building, going on. For many Irish towns it's the story of the last ten years, but one which has not yet been fully explored in television or theatre.

Currently, it is urban offerings such as *Fair City*, the Dublin-based soap which first went on air in 1989, and series such as *The Clinic* that have proved to be successful returnable dramas for RTÉ. Rural-based drama is confined to comedies such as *Killinaskully* and one-off productions. *Townlands* and *Vets on Call* provide well-made documentary tales of lives outside towns. Often beautifully shot, evocative pieces, they represent television telling rural life as it is, warts and all. For twelve years, *Ear to the Ground*, RTÉ's longest-running factual series, has provided those still employed in farming with a television product that represents issues that affect them economically and culturally and provides one of the last televised windows to their way of living. *Farm News* on RTÉ Radio One (currently broadcast daily in the evening *Drivetime* slot) and *Farmweek*, the half-hour weekend programme, provide current affairs for those still involved in farming.

Those of us who watched the box in the 1980s will remember *Mart and Market* and *Landmark* and their confusing lexicon of strange words revolving around the price of hoggets and finished weight of store bullocks. Online and print media such as the *Irish Farmers' Journal* now supply this sort of specific information to farmers; television is too broad a medium to cater for an increasingly specialist type of news information in a sector which has moved from mixed farms to very specific types of food production.

But if those who farm in a rural economy have their information needs catered for, is there a need for their lives to be represented in the arts, by writing, or television, or theatre? The obvious question to many in the broadcast media is: why make television or theatre drama exclusively about farming or rural life when the number of people involved in full-time farming is dwindling? Does it matter if we have no living pictures of Irish rural culture any more? Is rural culture itself deceased? Was it any coincidence that the best-known dramas set west of the Shannon in the last ten years were by a writer born in Camberwell, London? Martin McDonagh, whose darkly comedic plays have won him a Tony Award and international acclaim, has been likened to Synge by many, and his work references much that came before him. In 2007, the wheel came full circle when Roddy Doyle reimagined *The Playboy of the Western World* with Christy Mahon as an immigrant Nigerian.

If we in Ireland now see ourselves as international, as no longer unique, as part of a larger Western, consumer-based culture, does it matter? If the television and the media that we consume in Ireland is a picture of ourselves, we could say that rural Ireland and rural culture are dead and gone. The only place it's visible is in an almost exclusively Irish language context on TG4, in sport through Gaelic games and in the traditional music and arts that make it into mainstream programmes from time to time. There are many versions of rural Ireland, but a stranger who arrives in Ireland won't find much representation of it in either the Sunday newspapers or our national television channels. You have to go further afield to experience rural Ireland. The question is, how much further do you go before you fall off the edge of the map?

Chapter 7 ∾

| CONSUMING OURSELVES

Suzanne reflects on a Fendi bag

In 1997, through a combination of joblessness and wanderlust, Philip and I ended up living in Tokyo. The plan was to earn some valuable yen that would enable us to travel Asia for a year, bum around on beaches and flitter away our young lives in a climate of guaranteed sunshine. Which was pretty much exactly what we did.

Tokyo at that time was a city recovering from its 'bubble economy' crash, which saw Japan decline from one of the world's biggest economies to one of spiralling depression and negative growth. Yet even in those times of less plenty, it was a city of startling wealth and one of the world's best examples of an advanced urban community. With a population nearing 12 million, Tokyo's light-filled, *Blade Runner*-type cityscape was a place of phenomenal beauty; skyscrapers towered over city streets, motorways hung suspended over residential areas, and at the base of its tallest buildings were narrow alleys with wooden dwellings and bath houses, red paper lanterns still fluttering as the traditional markers of Japanese life.

In Tokyo your train is never late, despite the fact that the equivalent of the entire population of Ireland went through our local metropolitan train station, Shinjuku, every day. Crime is minimal and everybody is polite. On the surface it is a society which is ordered, wealthy and civilised, with huge attention paid to details of correct behaviour, dress and social form. A highly organised and sophisticated society, Japan recovered from the disaster of the Second World War and in the second half of the twentieth century built itself into one of the world's most successful economies, reaching a peak in the 1980s and early 1990s, when an uncontrollable property boom and a tidal wave of conspicuous consumption took hold. In 1998, the economy was on a downslide; housing had gone into super negative equity and the jobless numbers spiralled. Yet the consumption of designer goods—clothing, bags, shoes, skateboards, anything with a brand name plastered across it—thrived. Sound familiar?

In Japan, the cult of the branded product is absolute. No matter what your personal circumstances, it is deemed normal in Tokyo to wear luxury brands and in some cases to be literally covered from head to toe in designer logos. What looked like the strangest behaviour to us living there in the late nineties is now fairly mainstream in Dublin; women in department stores paying hundreds, even thousands of euros for a Louis Vuitton or Fendi bag, widespread panic at needing to have the new shoes, the new bag, the new haircut that will somehow define you as fashionable, rich and the envy of others. In 1997, Tokyo's obsession with the latest designer goods had reached unattractive levels of madness. Schoolgirls (all of whom carried black Prada bags—that year's fad) were known to prostitute themselves in order to earn money to buy more designer goods. Initially we didn't believe this story, but it was confirmed by our Japanese peers, whose reaction was 'yeah, loads of them do it'.

Even odder, and sadder, was the rumour that many women in their twenties, some married, kept their school uniform washed and ready to go for a bit of income on the side, income that was again largely used in the local branch of Vuitton or Chanel. The school uniform is a cult item in Japan. It is sexually charged and a staple of porn magazines, porn comics, and both adult and youth culture television and film. The sexualisation of extremely young people and children was also fairly common: alongside their awareness of themselves as sexual commodities, teenagers knew that they could use this value to further their own social value by covering themselves head to toe in designer goods.

Okay, close your eyes and open them again. It's the wet summer of 2008 and we're in Dundrum Town Centre in Dublin. It's a weekday and the shopping mall is full of school-uniformed girls with a selection of fake and real designer bags dangling from their elbows. Some of them have just left Starbucks, where they paid €4 for a mocha, which they are carrying in a cardboard cup and pretending to drink. After all, they could do without the 350 calories inside, especially if they are going to fit into those Victoria Beckham jeans. And their mobile also needs updating; I mean, if you can't take decent pictures of your friends, how are you meant to look popular on Bebo? This is a snapshot of some of Dublin's youth and the kind of concerns that keep them awake at night. But it is not only them who are populating the shopping mall, it's a whole load of other people: us. Mothers, husbands, grandparents, students, shift workers, taxi drivers, all shopping. Why did shopping become Ireland's biggest leisure activity? When did it begin to define who we are? How did we all become so enmeshed in consumer culture that we literally turned into a bunch of shoppers?

GREEN LUXURY

This isn't just happening in Dublin—all over the country a rash of malls, shopping meccas and outlet villages have opened over the past five years. We can hope that the full madness of Tokyo's conspicuous consumption remains far from our shores, but many aspects of extreme consumerism thrived and became enmeshed in Irish society during recent times. What was so surprising was how rapidly and successfully Irish people fell under the spell of luxury brands and very highly priced consumer goods. But why? We thought of ourselves as a county with its head screwed on, so what made us dive so far down the path of rampant consumerism?

The laughable thing is that in Ireland, as recently as twenty years ago, it was considered vulgar to wear your money on your sleeve. How have we come to do the exact opposite, and what kind of a toll does this take both on our own society and on the system that produces the never-ending supply of goods that we think we need? Does coveting designer goods that we think will make us happy bring with it the other less attractive hallmarks of life as it was in Tokyo: social alienation, fractured traditional community and family values, status-related peer pressure, *very* high rates of suicide among both adults and young people? Is this where we are heading but just don't know it yet?

It's also shocking how quickly we embraced, in particular, the really expensive stuff at the ludicrous end of the scale; the €400 jeans, the €500 shoes, the bag that cost over a grand. Yes, it's crazy stuff, but how many of us have just gone out and spent this kind of money on clothes or accessories without even thinking? The truth is, over the past ten years we Paddies fell hook, line and sinker for luxury lifestyles. The luxury goods industry is worth $157 billion worldwide. It has many detractors, mostly those who point out that by buying this stuff we are benefiting no one but the huge multinational groups such as LVMH (Moët Hennessy–Louis Vuitton), which profit from the branded perfumes, bags, polo shirts or whatever that we buy from their vast array of product lines. Many of us might kid ourselves that we don't fall for designer goods, but take a look at the products on your bathroom shelf or the perfume in your bag—there are almost certainly some luxury brand names lurking there, and all of us have fallen for it.

Despite the allure of advertising, which suggests that these design houses are producing covetable artisan hand-made goods, the reality is fairly different. These are globalised common or garden luxury goods, churned out in mega quantities more akin to the production of a McDonald's burger. In terms of clothing alone, the LVMH conglomerate owns Fendi, Louis Vuitton, Celine, Donna Karan, Pucci, Mark Jacobs, Givenchy, Kenzo, Loewe and Thomas Pink. The implication that Donna Karan herself or Monsieur Givenchy (now long dead) oversees every lovingly stitched detail of that €800 dress is pretty far from the truth. These items are made in factories located for the most part in the developing world, and the perception that luxury brands are of higher quality, and therefore more desirable, is often a fallacy.

Recently, American journalist Dana Thomas lifted the lid on many of the dirty secrets that thrive in the luxury goods trade. She travelled to Guangdong Province in China to investigate her suspicions that many luxury goods are made in far from luxury conditions.

> Yes, luxury handbags are made in China. Top brands. Brands that you carry. Brands that deny outright that their bags are made in China make their bags in China, not in Italy, not in France, not in the United Kingdom. I held them in my hands. To see them, I had to promise the manufacturer that I wouldn't reveal brand names.[1]

Furthermore, in the factories that she visited, the brands made the manufacturer sign a confidentiality agreement stipulating that he could not reveal the fact that their hugely priced bags were produced not in Italy, but in China.

No wonder the brands make money out of, in many cases, duped consumers. Most of the top-name handbags retail at ten to twelve times their production cost. What began as artisan family-name suppliers to the French courts of the eighteenth and nineteenth centuries have become global names churning out everything from perfumes to flip-flops. For the big fashion houses, couture is a loss maker, and they now rely on the products that any regular plebeian can afford. You may not have the cash for a €200,000 couture Dior dress, but the sunglasses could well be within your budget.

THE REAL PRICE OF LUXURY

Anyone born in Ireland before 1960 is prone to making out that they lived a life akin to child enslavement in a Lancashire coal mine of the 1750s. Parents and grandparents complain that they 'just never understood the new Ireland'. 'The money . . .' (shaking their heads dramatically), 'the cars ... the shopping morning, noon and night.' In their day a carrot was a yearly treat and entertainment was catching mice for the Christmas dinner. Conspicuous consumption, or showing off one's wealth, was strictly outlawed, and reserved only for the mad. In culchie Ireland you kept your wealth (if you had it) to yourself. No one should ever know what you're worth. Perhaps if they did, they might try to take it off you. Assets were secrets and wealth was disguised with frayed cuffs and battered donkey carts. What a contrast. How did we go so quickly from a culture that hated anyone knowing what money you might have to one where we can't tell enough people how rich we are? We even have annual lists to prove it. Don't know what I'm worth? Read about me in the *Sunday Times*.

But even as we were being told again and again to consume more, it turned out that the rise in luxury goods and our lust for shopping were neither sustainable in an environmental context nor fulfilling in terms of our own personal lives. Keeping up with the Joneses is a particularly strong force in Ireland. As the economy over the last ten years got richer, we all wanted to look the part, whether we had the money or not. From being a sceptical, keep your cards close to your chest sort of nation, Ireland flew headlong into a spending mania. The lauding of the rich and well dressed in newspapers and the social pages of magazines also promoted the message that this was the way to look and live. It told Irish people that this was the direction we should be going, and luxury goods envy began, creating paranoia among those who could afford to keep up with the latest clothing and accessories fashions. Some find themselves in a state of sheer panic if they can't get their hands on what *Vogue* is telling them is this season's absolute must-wear. It's an endless cycle designed precisely never to fully satisfy the consumer, for when you've reached number one on the waiting list for your Hermès Kelly, the word in the glossies is that the classic Chanel quilted bag is the one to have, and it's 'Hey! Let's start all over again.'

For those of us who can't afford what's hanging on the arm of this month's celebrity, at least there's the high street; the cheap as chips clothing alternative. 'Thank God!' we say, breathing a sigh of relief. But let's look at this a bit closer. Nearly all the high street shops, whether they sell clothing, electrical goods or DVDs, are part of English or internationally owned conglomerates, such as the Arcadia group owned by the gazillionaire Philip Green. We also know somewhere deep inside that when we buy really low-priced high street stuff, there must be some kind of price to pay, whether in terms of the way these products are made or what the people who produce them are paid. We are happy to buy Primark products, but we certainly wouldn't be happy to earn what their factory workers take home in India.

Ireland's consumer binge of the last few years has made people like Philip Green very rich indeed and prone to partying big style. He spent £4 million on his son's bar mitzvah, when two hundred guests partied to music from Andrea Bocelli and Destiny's Child. For Green's fiftieth birthday he flew guests in a chartered 747 to Cyprus for a three-day toga party where they were serenaded by the snakehip kings themselves, Tom Jones and Rod Stewart, who were reportedly paid £750,000 for a forty-five-minute set. It's no surprise that Green has been described as 'flash'.

But against the backdrop of all this partying came the allegations in 2005, by organisations such as Labour Behind the Label and No Sweat, that Green's clothing lines were employing sweatshop labour. In the 1980s and early 1990s, no one thought twice about where the clothes they purchased originated from, and many were surprised when campaigns aimed at Gap and Nike revealed that many workers in Asia and the developing world were suffering from our fashion demands. Western companies are now more aware of the importance of producing clothes more ethically and of the disastrous publicity that ensues when they are found out not to be toeing the line.

However, in the developing world, where textile work is often subcontracted out by the main approved contractor, 'slave' labour can still be very much a reality. Just last year, Indian children as young as ten were found working in a factory in conditions described as close to slavery to produce clothes for Gap Kids. The children described long hours of unwaged work, as well as threats and beatings. Gap said it was unaware of the conditions of production as the contract had been subcontracted without its knowledge to sweatshop labour, and withdrew the garments while it investigated the circumstances of the breaches in the ethical code it now imposes. With celebrities such as Madonna and Sarah Jessica Parker endorsing its products, last year Gap launched a huge advertising campaign for Product Red, the charitable trust founded by Bono.

As our demand for cheap clothing subsumes our willingness to understand how it was produced, it is likely that there will be continued breaches in ethical production codes such as Gap's, or, more likely, wholesale abuse of workers where companies far below the publicity radar turn a blind eye to how their clothing is being produced.

Professor Sheotaj Singh, co-founder of the DSV, a Delhi-based centre for rescued child workers, says he believes that as long as we buy cheap clothing sold in high street stores, there will be problems with unscrupulous producers who will employ low-cost labour or children: 'Consumers in the West should not only be demanding answers from retailers as to how goods are produced but looking deep within themselves at how they spend their money.'[2]

And to prove that those who stick their heads above the parapet in campaigning for workers' rights sometimes risk their own lives, in March 2007 a trade union organiser in the Philippines was gunned down at eight o'clock in the morning; in December 2006, his predecessor had been killed at the factory gate. No one has been apprehended for either murder. Big business speaks, and the conditions in which our clothing is produced is certainly worth thinking about in terms of the increasing volumes of it that we buy, and where our money is going.

BUY IT, USE IT, DUMP IT

Ireland is now a culture that embraces disposability: cheap clothing, cheap food and items that provide us with an instant shopping 'hit' but are soon out of date, underused, no longer in fashion or never worn. The cheaper clothing in particular becomes, the more of it we seem to buy, loading up our baskets as if we've won the Lotto. 'A hair band for fifty cents! A bikini for €12!' We can't get enough of the stuff. Yet we seldom stop to think that if you can buy a scarf in a supermarket clothing section for €3, how much was the person paid on the other side of the world to make that scarf? Under what conditions were the materials to make it grown? Does that matter? If it costs €3 and you walk away with a bargain, are its origins in any way important?

The decreasing cost of things we purchase also means our attachment to these items are minimal. How many times have our shopping sprees met an ignominious end in a pile of unopened bags sitting in the hallway of our house? The excitement of the purchases was extinguished the minute you placed them in your car and made the journey home. Once you've opened the door of your house and wearily dumped the bags on the floor, the shopping buzz is a distant memory. Not only that, but the task of unpacking it all, and disposing of or recycling the ridiculous amounts of wrapping that every purchase now seems to drag with it, adds to the distaste for dealing with the entire lot. Make a cup of tea instead: *Coronation Street* is just about to start.

Apparently, unopened shopping bags and lack of interest in the purchase once you get it home is one of the hallmarks of shopping addiction. 'But I'm not a shopping addict,' you shout. 'Okay, ummm … maybe a few cheap outfits here and there, a new gadget every now and then, a few more DVDs and CDs a month.' Shopping these days is not about need but about filling an emotional hole. You're overtired, you're emotionally unhinged, your boss is getting on your nerves: you need a reward. Go shopping. Buy that Fendi wallet; you're worth it. Buy that seventy-inch flatscreen; after all, television is your biggest pleasure in life. Other handy emotional-hole fillers are bottles of Australian Chardonnay, lines of coke, huge slabs of chocolate and large packets of Kettle chips from the local Statoil.

During the boom years in Ireland, we went shopping mad. Was it sometimes the case that we bought stuff because we were unhappy? Were our lives in Ireland half-empty rather than half-full? A poll conducted by the *Irish Times* in 2008 found the emotional state of men in this country generally very happy. 'The conclusion thus far has been that despite the current economic turmoil, men have maintained an optimistic view of life in general, seeming to take a more pragmatic view of the negatives within the grand scheme of things.'[3]

The *Times* concluded that the typical Irish man seemed reasonably content and at ease with himself, despite the fact that eight per cent of the survey's respondents said they had 'seriously considered suicide'. Not exactly great news, that part, is it?

Worldwide, our 'greed is good' attitude and overconsumption throughout the last decade have led many of us towards the disease of our age: affluenza, a preoccupation with earning more, spending more and ultimately wanting more. Unfortunately this kind of thing doesn't ultimately make us any happier. Psychologist Oliver James says that defining ourselves through our earnings, possessions and appearances is making us more miserable than ever. He travelled through seven countries and interviewed hundreds of people in order to pin down the roots of why we buy so much stuff. Among middle-class consumers, he found increasing incidence of mental illness, depression and unhappiness. He also found that across English-speaking countries, consumerism had led to a type of sameness and the erosion of cultural differences.

Worst of all is the feeling of homogeneity. There is little room for eccentricity, for individuality. Individuality has been replaced by consumerism. People have confused the idea that they are expressing themselves as individuals with the idea of purchasing goods and services.[4]

Perhaps less dramatic than being mentally ill or completely culturally empty, many people in the Ireland of the past decade shopped just because they were bored and had nothing else to do on a Saturday but get into the car, drive to the local mall (often several hours away) and go shopping. To most of our parents' generation, this behaviour represents a kind of Americanised crazed consumerism, the likes of which they saw on that once-in-a-lifetime holiday to Florida in 1984 and agreed that 'no, it could never happen in Ireland; people are far too sensible. Aren't they?' Well, now it's here, and they cover their eyes in disbelief. 'Haven't you got enough clothes already?' they say. 'Why don't you spend some time cleaning up your garden, or visiting Aunty Margaret down the road?' Yeah, you're right Mum. But I still really want that Ralph Lauren duvet cover …

The problem with a shopping habit is that it's a bit like a crack cocaine habit; as time goes on, we need more and more of the stuff to create the same emotional high that we used to get from it in the virgin days when we began the habit. Due to the cheap credit of the boom years, shopping became an occupation in itself, something you did on a whim, giving an immediate high which dissipated pretty quickly. This is very different from how generations before us made their purchasing choices. They saved, they waited and when they had enough money they made the purchase; and no doubt this earned a satisfaction far greater than most of us feel today. Like that feeling of going to the local shop when you were eight years old: you had ten pence in your grubby hand, a nose pressed up against the window admiring a myriad of purchasing choices; twenty half-penny cola bottles or two flogs? One packet of Taytos or five gobstoppers? The pressure of choosing was immense. If something has to be saved for, waited for, lusted over, its return for the owner when they finally get hold of the item is a much richer one than something that's easily obtained. How many of us have clothes from the well-known high street chains that were never worn, finally thrown into bin bags and given to Oxfam? (Thank God Oxfam is there to salve our greedy, First World consciences.) Otherwise, our over-shopping might just make us feel a teensy weensy bit, well … bad.

'But if I'm not dressed well I won't fit in', we think. 'My clothes are out of date, I must have the new season bag, my ten other pairs of black trousers just won't cut it this winter.' It's worth remembering that magazine editors are paid precisely to tell us to constantly renew our wardrobes and purchase endless streams of cosmetics even though we might have ten pink lipsticks already. It's not that the editors themselves crucially believe you need the New

Bag to be acceptably attired this season. They desperately need to sell the ad space in their magazines to fashion and beauty retailers—otherwise they'll be out of a job and will have to sell their seventeenth-century bastide house in the Dordogne. Poor them. Some will even admit to the artificiality of the never-ending retail merry-go-round: it's fall—the equestrian look! Spring—the maritime look! How many times can plaid make a comeback? Constantly selling us the same ideas in a different way keeps the multinational clothing groups afloat. They have to make us consumers believe we are simply not good enough *without* their products in order to turn a profit. It's a cycle which at its heart feeds on our—particularly women's—insecurities and our need to feel accepted and rich, and thus to gain status from those two things.

The problem is that even if our wardrobes are overflowing, we live in a culture that constantly tells us that we need Stuff. Stuff will make you happy. Stuff is what makes the world go round. You *need* that Paul Smith coat, the Neff oven, the Louboutin shoes, the Aston Martin Vanquish (the five litre or the 6.2? Feck oil prices, they can't go up forever, can they?). We are told that if we buy what our hearts desire, our cup of happiness will overflow. The wealthier a society gets, the happier it should be. Isn't that right?

Unfortunately, no. Research has shown that when people have their basic needs—food and shelter—met and have access to education and a belief in upward mobility, that's when they are at their happiest. This probably occurred in Ireland in the decades between the 1960s and the 1980s. Things were looking good then; most families during that period had enough means to live pretty healthy and (by world standards) well-upholstered lives. When it starts to go downhill is when societies have too much wealth and too much choice. Rampant consumerism, the break-up of family and community, all the sort of stuff that Tokyo was exhibiting in the late 1990s, is going to be increasingly familiar territory in the Ireland of the twenty-first century.

The problem with having consumerism as our main activity, almost a *way of being*, is that it will give you a high, even if only for a little while. Yes, you can look lovingly at your BMW and imagine that it is looking lovingly back at you, but at some stage your mobile phone will ring and you are back in the real world, working hard in your job, sitting up with sick children, trying not to row with your spouse. Just when you've constructed the fantasy ad picture of the beautiful home, the Marc Jacobs outfit, Real Life intervenes in all its pesky horror. Is this why so many people in Ireland today claim they feel stressed and under pressure? Because we are told that if we look the right way and buy the right stuff then our lives will be good—we are rich, we are thin, we go to the right restaurants, we should be happy. Then we find it doesn't always work out that way, and the coping mechanisms our parents had (patience, prayer, belief that life was hard and if anything good ever happened it was a goddam miracle) are skills that we have to dig deep to find, if we can find them at all.

Recession has finally put the brakes, for the moment, on our shopping mania. Retail figures have fallen dramatically as Irish consumers are opting to limit their spending. Much of this spending over the boom years was money that many of us didn't really have in the first place. For the past decade or so, rising house prices made people feel wealthy irrespective of their weekly take-home pay. And so shop they did. Now that the R word is biting hard on our wallets we've calmed down, the malls are quieter, some shops are closing and most people report that they are shopping less and watching what they spend when they do shop a hell of a lot more than they did a year or two ago.

Just when everything began to look more realistic, in 2008 England built the mother ship of all shopping malls, Westfield, in the centre of London, which is now Europe's biggest city-centre shopping mall. It's been called the Cathedral of Consumerism and cost £1.6 *billion* to build. Obviously its investors are relying on people not losing their shopping lust anytime soon. But even in London, one of the richest cities in the world, many people are expressing shopping fatigue. Writer India Knight said of the opening of the Westfield mall:

> I love shopping so much that I once wrote a book about it. But even the most ardent devotee can't help but notice that shopping as we have understood it for the past two decades, from the more-is-more 1980s onwards, is losing its gloss. This year, the cool women I know are knitting and making chutneys to give as Christmas presents … witness the return of hand-me-downs for children, the longer queues at shoe repair booths and the fact that everyone I know seems to be buying a bicycle.[5]

Gosh, what's happening? Could shopping be losing its popularity? One important trend Knight identifies is that the decline in shopping is not simply because people are more strapped for cash. It may also stem from the fact that shopping as an activity in itself is no longer as fashionable as it was. Loading the SUV with masses of designer bags is viewed as a sign of tacky WAG-type behaviour in some quarters. Making one's own and cutting down on shopping is a trend that some people are choosing rather than being forced into. This New Thrift is, of course, not new at all; it's how most people lived in Ireland until very recently.

GYM? WHY DON'T YOU BLOODY WELL WALK?

During the good times, older people in Ireland could be annoyingly carping about our newfound wealth. They rained on our parade, burst our bubble and reminded us constantly that it would all come to one big disastrous end. Many were delighted when the downturn in the economy finally hit. 'I told you so!' they cried. 'Real life is intervening at last! Back to sackcloth and ashes!' While we were crying into our cocktails, they told us to get real, be practical, stop shopping and believe in something; have faith that human existence is not just an arbitrary shipping passage on which you need the best-appointed cabin. Do something of worth, be self-sacrificing, sometimes even poor. You might begin to value what you have.

The changes in Ireland over the past thirty years have altered society in a way that many older people find confusing and sad. They see empty churches, overworked families, children being cared for in crèches by strangers, neighbours dead without anyone knowing for days. They see high rates of divorce and suicide, particularly among young people, a reliance on electronic communication instead of meeting and spending time with people face to face. As a richer community, have we in reality become a poorer one? The way we live now as people and communities is profoundly different from how our parents grew up. They find themselves having to adjust to a newer Ireland, and perhaps we sometimes find ourselves having to listen to what they have to say: that money isn't everything, that it's the small things that make you happy. Working eighty hours a week and buying expensive stuff just doesn't do the trick.

If affluenza and greed across the Western world contributed to economic near-collapse, can we learn from the past? Were older people really better off in the times they grew up in? Is getting back to basics a plausible reality or a heavily rose-tinted dream? Were people's lives really better than they are today? Certainly not in terms of the basic stuff like health care, diet, life expectancy and child mortality. But were people happier? There was no statistical analysis of happiness in the 1930s. Even if it had existed, the interrogation of one's own state of mind would have been unusual to most people. Many would have answered that they were too busy working and too physically tired to think of their own happiness. Happiness was for heaven, and if you lived a good life, and prayed, you might eventually get there.

The negative aspects of the Catholic Church—smothering personal joy and whittling people into a system of repression and emotional inertia—have been well argued elsewhere. Yet most older people claim the Church had a stabilising influence on society, promoting kindliness, community involvement and a positive motivation towards others that is sorely lacking in modern times. As churchgoing populations fall, what has replaced, or will replace, the Church as a system of belief or blueprint for living in Ireland? Does it matter? Many Catholic-born Irish people feel that if their expensive, militarily planned wedding features a church and their subsequent children get baptised inside one, that's enough. Conscience salved. 'You don't have to go there every Sunday, for God's sake. That's just taking things too far.' Does walking past the doors of the church mean that we no longer believe in anything? Or is it that we believe very confidently in the ability of ourselves alone to deal with the problems we encounter in life? In modern Ireland, it also helps that there is an army of therapists, counsellors and behavioural problem-solvers at whom we can throw money to sort things out. 'Why would you go to church when you can get professional help?' Is the Church now completely irrelevant? Have we as a thoroughly modernised, Tokyo-ised society ditched religion along with the stuff of the fairies?

Even if we are no longer watching the Church, it might come as a surprise that it is watching us.

THE DEVIL IS PRADA

The Vatican itself, in a bid to address our out-of-control consumption habits, brought its list of seven deadly sins up to date by adding seven new ones for the age of globalisation. The new list was published as the Pope deplored the 'decreasing sense of sin' in today's 'secularised world'. Bishop Gianfranco Girotti, head of the Apostolic Penitentiary, the Vatican body that oversees confessions and plenary indulgences (a barrel of laughs, no doubt), said that priests must take account of 'new sins which have appeared on the horizon of humanity as a corollary of the unstoppable process of globalisation'. Whereas in the past sin was thought of as an individual matter, now it has 'social resonance'. 'You offend God not only by stealing, blaspheming or coveting your neighbour's wife, but also by ruining the environment, dealing in drugs, and social injustice which causes poverty or the excessive accumulation of wealth by a few.' He said that hedonism and consumerism had even invaded 'the bosom of the Church itself, deeply undermining the Christian faith from within, and undermining the lifestyle and daily behaviour of believers'.[6]

While indicating that trading in the temple is still very much out of favour, the publication of the new list of sins coincided with Pope Benedict's visit to the United States, where Pope on a Rope, Benedict Beer and other such articles of rampant commercialism headed the tidal wave of Pope memorabilia for Christians interested in joining the party. In the interests of good taste (and no doubt aware of the new sins list), the Archdiocese of Washington asked that the plastic bobble-headed Pope doll be withdrawn from an ad publicising the Washington Metro system. It seems that bad taste has its own particular limits.

So, bobble-headed Pope dolls aside, if we listen to our parents and make noises about trying to get back to basics, does this mean we have to turn back to the Church? To live simply and more happily, must we give up everything we have come to take for granted in gold-card Ireland? It's become clear in recent years that if the Church isn't enough to frighten you into living differently, there's always the threat of being flooded out of house and home by the effects of climate change. If the Vatican is telling us that the Devil *is* Prada, it's worth remembering that everything we consume, buy and subsequently dump has a carbon footprint and a nasty manufacturing cost that, whether sinful or not, makes much of how we live in modern Ireland completely unsustainable.

Is the Catholic Church pre-empting the costs of living as a community in a globalised world? Is our over-shopping and over-consuming making us fat, removed from reality and depleting both our bank balances and the planet's threatened resources? Even the small things matter at this late stage in the game: every plastic biro we buy has used oil as part of its production process, let alone every car we buy, every long-haul trip to the Maldives or cheap Ryanair flight to the south of France. What we use from the biosphere is limited. When we put an economic value on what the environment can produce, it changes the way we use it and risks the very obvious path of depletion, which is where we're heading now. Is the accelerated rate at which we now consume stuff a shortcut to the end of our own existence?

Generations ago, our parents ate food produced largely within their local area. Services, crafts and most items in their household came from a production process that was under a day's journey away. Aside from tea, sugar and a few newfangled foods from hot countries around the world, they consumed goods and services that were produced in a simple and sustainable way. Things were mended when they wore out, leather was patched, holes were darned and the basic skills of sewing, carpentry and, most important, growing your own food were taken for granted. Crafts and skills common even twenty years ago are now very thin on the ground. We no longer fix stuff, we throw it away into a big hole in the ground. Or we ship our stuff off to China, to be separated into different bags of stuff which are then thrown into big holes in the ground, or sometimes, if we're lucky, recycled into other stuff which we then go out and buy.

Perhaps recession, climate change and our voracious, damaging levels of consumption will help change our habits, make us ease off on the spending and take time to savour a slower life. Do you really *need* that new car to make you happy?

In 2009, fashion commentators cried into their champagne and claimed the It Bag was dead—basically, too many people have one. The democratisation of high fashion has also been its downfall; the new ubiquity of brand logos and their Bangkok rip-offs have sent some people running in the opposite direction. In a sense, making something accessible to common working people devalues it for those with more money in their pockets. Editors are now telling us to buy vintage, be thrifty and swap clothes with friends. But alongside their green, 'The New Downsizing!' headlines, they are still selling us big brand clothes and pushing ads for the giant luxury goods groups. Magazines must make their margins too.

Getting back to basics might be a good thing, but living the life of an Irish farming family of the 1940s may not be everyone's ticket. Yet somewhere between the way we live now and the way our parents lived, there has to be a happier, more enriched life with a more sustainable outcome. For those of us who claim we can't slow down, it's usually because we don't choose to. Changing the way we live and consume in Ireland is a choice, but it's something that may be forced on us sooner than we'd like. We may not quite go back to the donkey and cart, but peak oil prices and the limiting of motor transport will become realities in the next twenty years. Flying potatoes from Israel to your local supermarket might seem a little bit extravagant when you consider how easy it is to grow your own. How we choose to live in this future Ireland is simply up to us.

Can Fendi bags make you happy? I don't think so. I own one, and I don't even know where it is.

Chapter 8 ∾

׀ GREEN IS THE NEW BLACK

Leading us up the cottage garden path

Cameron Diaz is driving a hybrid car. Croke Park is going carbon neutral. Your child or spouse has turned themselves into a Reichsmarschall of home recycling.

Scary people tell us that if we don't turn into greenies, we will die. If we don't change our wicked ways, we will be flooded out of it and be thrashing around in lifejackets. In the time it takes Jeremy Clarkson to say 'six-litre engine', we will have killed off several species of the planet's animal life. The polar bears will come after us with hatchets, Spar will run out of apple-flavoured bottled water and it's all going terribly wrong.

What's it all about? Green is the new black, whether we like it or not.

At the frontline of this crusade we have the green fashionistas, the Duncan Stewarts, Hugh Fearnley-Whittingstalls and Gwyneth Paltrows of this world. Their plan of action is to charm us into being green, to lead us down a cottage garden path where being green is good, a code for living born out of quality of life, not fear. Their green is soul-cleansing and healthy, something that augments rather than curtails your life. It is a return to the past by way of the future. If you are green, you are good and you will go to heaven, not rot in hell for buying supermarket pizzas and driving an suv.

In recent years this growth of green as fashion has caught many by surprise. In the beginning, environmentalists had beards, supposedly ate lentils and would not be fun to invite to your Christmas drinks party. But suddenly it all changed. Green living is now appealing, desirable and, in some cases, competitive. But is this just a fashion moment, a change in attitude and behaviour bred from alterations to our lives that we have been unwillingly forced to make? In Ireland, are we really committed to looking after our environment, sourcing ethical food and reducing our carbon impact, or are we paying lip-service to a movement that we treat as a fashionable blip? Most of the changes that Irish homes have made to become greener have come from forced rather than optional choices. Most of us now recycle our rubbish, as waste charges became a factor limiting the amount we could afford to dump in landfills. Some of us now switch off the lights a bit more frequently and have turned down the thermostat on the central heating because the last bill was so big you thought you were reading the wrong figures.

It also helped that in Ireland in recent years we have seen contaminated water supplies, flooding and salmonella in our food plants. Perhaps the reason being green has segued into being fashionable is that we have had to make do with situations that are forced on us and to turn them into something positive; a bit like girls in wartime Britain painting black lines on their legs to imitate stocking seams. What's born out of necessity often becomes ordinary and then, by extension, attractive. Before you know it, you're at the checkout in Avoca buying organic flour to make your own bread and a few seeds to grow lettuce on your windowsill. It starts to get in on you. You start to believe it.

KEIRA KNIGHTLEY'S BAG

One good example of the green movement's transformation from boring to high fashion came with the advent of the reusable shopping bag. Ireland introduced its plastic bag tax in 2002; banning supermarkets from giving them out free to customers inadvertently created the age of the reusable bag. Having begun life as a pretty ordinary item, the reusable bag has embarked on a fantastical journey of identity crisis, from something the dog lies on in the back of the car to an item retailing on the shelves in swanky stores like Harvey Nichols and Neiman Marcus in New York.

The first reusable bags were tough plastic things sold in supermarkets and branded by Tesco, Superquinn, etc. But soon label-conscious shoppers and manufacturers (never shy of a commercial opportunity) cottoned on, and the reusable bag began its journey from something very ordinary to the pages of French *Vogue*. Bookshops, convenience stores and every kind of eco-nonsense brand under the sun began producing a bag under their name, and consumers were only too happy to get in on the act. After all, why not advertise our green credentials on our arm, even if the contents within are two tubes of Pringles and a giant bottle of Coke?

The green companies were soon followed by eagle-eyed luxury goods conglomerates, with Mark Jacobs (his canvas a snip at €165) and Yves Saint

Laurent launching their own reusable grocery bags. The trend reached Evel Knievel heights of madness when designer Anya Hindmarch's square cotton shopping bag labelled 'I am not a plastic bag' took the fashion world by storm in 2007. Once celebrities such as Keira Knightley were photographed carrying the bag (which cost a fiver), ordinary punters fell for it hook, line and sinker, and there were waiting lists—yes, waiting lists—in London and Dublin for a bag to carry your groceries in.

What's interesting about the reusable bag's popularity is how it became a symbol for those who wanted to be *seen* to be green. But why is being seen to be green suddenly important? Because being green is now a fashionable, desirable quality. It's the Chris Martin and Gwyneth Paltrow way to live your life. How it came to be viewed in this way is a curious phenomenon, but it has much to do with our association of green ideas with good-quality-of-life ideas. Again, this is often celebrity led; if the people we regard as high status in society—actresses, rock bands, TV presenters, celebrity chefs and sometimes, yes, sometimes even politicians—are telling us that one particular code of living is a desirable one, it's likely that we will change our attitude to seeing green as a positive rather than a boring quality. The medium is the message, they say, and if Cameron Diaz is your medium, you're much more likely to pay attention to her than to Trevor Sargent talking about low carbon emission engines.

BUT OF COURSE WE PEE IN OUR GARDEN!
While Hollywood A-listers yak on about how green they are, at the extreme end of the scale are the carborexics; those greenies so dedicated to the cause that they inflict huge life changes on themselves (and often others) to live with as little impact as possible on the environment. It's the extreme end of the scale, but already in Ireland there are those of us adopting a carborexic way of living. Behaviours that many would find extreme are normal routine for the carborexic; living with little or no central heating, weighing household waste, foraging for wood and food, eschewing fridges and other energy-eating appliances, never using a car, and so on. A report in the *New York Times* found evidence of all manner of new lifestyle choices in the US that might be considered carborexic or 'dark green' activities.[1] These included buying only second-hand consumer goods, having your children sleep together in one bed to save on heating bills and using your lawn as a bathroom to save flushing the toilet.

Nice. But before we fall about laughing, there are plenty of people in

Ireland already dedicating themselves to many kinds of energy-saving activities, low-fuss things such as reusing plastic bags, turning down thermostats, saving rainwater for the garden or using the half-flush option on their toilet. What makes these green-conscious individuals different from the carborexic is that the carborexic has become zealous to the point of obsession about their environmental impact. Some have likened the trend to a type of obsessive compulsive disorder, where behaviours are repeated and tend to grow more extreme as the person suffers severe anxiety if they fail to follow their particular 'routine'. This may start off with something simple such as turning off the lights in the house, and move into more time-consuming activities such as recycling every single piece of your domestic waste, eating only organic foods, etc. Many people who adopt green strategies in their households admit that green habits can start to grow on you in strange and unexpected ways. But when being green becomes an obsession, something is surely going a bit wrong in your brain.

'If you can't have something in your house that isn't green or organic, if you can't eat at a relative's house because they don't serve organic food, if you're criticizing friends because they're not living up to your standards of green, that's a problem.' So says Elizabeth Carll, a psychologist who is witnessing the growth of obsessive green behaviours in the us.[2] While initially we might have been forced to include green practices such as recycling household waste in our daily routine, 'greening' your life soon seeps into other spheres, and once you adopt new habits, such as composting and growing your own food, it becomes very hard to go back to your old wicked ways.

Wicked is the key word here. The very idea that to be un-green is to be somehow morally bad is another reason suggested for the crazy habits of those who adopt extreme-green lifestyles. Some observers have said that carborexics are people that in another age would have been attracted by evangelical or extreme religious tendencies. They associate being green with being morally good, and they see themselves as somehow essential in a campaign to save the planet. This begins to affect their relationships with other people, whom they judge on a scale of how green/good they are to how un-green/bad. An acquaintance who has given up all aircraft travel forces her husband, who doesn't share her evangelism, to do the same. This meant that a recent trip from Birmingham to Ireland to see his friends involved a trip on a ferry, which docked at six a.m., followed by a six-mile bike ride in the rain into Dublin to catch a train. His wife had asked him not to use a taxi for this leg of the journey, so he brought his bike over with him on the boat. Is it unfair of us to judge this as going to ridiculous lengths on behalf of someone else's beliefs, or is it just modern-day thrift and a use of alternative, simpler transport?

What we do know is that in Ireland we have already begun to green our lifestyles. What we don't know is how much we will alter our behaviour when it gets to making difficult, more uncomfortable choices about our environment and sustainability. Will our impulses to do the right thing stop when it gets really difficult? Like cutting aircraft travel out of our lives, or only eating food produced from our own garden? Growing your own food is already soaring in popularity, spawning a Good Life trend of home farmers all over the place. If you don't know your celeriac from your swede, or how to grow them in a well-composted plot outside your kitchen window, in some food circles you're an ignoramus of the highest order. Beware: the urban farmer is cropping up everywhere.

BUT I'M A *REAL* FARMER

Some blame Felicity Kendal and her role in their early sexual awakening. Others name a single incident, a food-themed road to Damascus moment when they discovered their inner farmer and decided to grow their own food. Perhaps it was coming across the Irish website meatyourchicken.com where they discovered how the average supermarket chicken went from egg to packaged meat in forty-two days. Or perhaps they fancied Kate Moss in her Hunter wellies, or Dolce&Gabbana's equestrian winter collection. Whatever the reason, grow-your-own enthusiasts are popping up everywhere, from allotments in Dublin 2 to chicken coops in the suburban garden and herb window boxes in Waterford apartment blocks.

Whatever the starting point, there is a boom in the number of Irish people growing their own produce and, in particular, in urban farmers. Urban farmers are city or suburbs based. They are a distinct bunch from hobby farmers, who tend to live in rural areas, and whose ambitions might encompass a couple of breeding heifers and a plot of land slightly larger than a living room rug. Hobby farmers often live close to real farmers. Real farmers are culchies who do glamorous jobs like cleaning scour off their milk parlour walls and leaning over shit-sprayed rails at marts. Real farmers think hobby farmers are certifiably crazy: who would undertake the drudgery of farming if you didn't have to make a living from the damn business? Yet these are the same real farmers who on retirement still walk the land every evening, casting an eye over the few remaining bullocks, just in case some of them might drop dead from too much condition. The fact remains that whether you're a Barbour-wearing urbanite or a real farmer, getting pleasure from doing a good job on something you grew or reared yourself is the same for all farmers, no matter how much land they have or how intensive their business.

For our urban farmer, it's all on a much smaller scale. Raising his own cattle is a distant dream, though he's in talks with a friend who has a large suburban garden about buying two Tamworth pigs next spring. It all began when a fashionable friend's table of impressive dinner party food apparently came from her tiny garden in Dublin 8. While half suspecting she might be fibbing (did she actually pull that chicken's neck herself?), the general consensus among the stylish and influential around the table was that growing your own was the only way to go. 'I mean, the price of veg these days, it's shocking, let alone the pesticides and rubbish sprayed on everything that isn't organic! No wonder we're all dying of cancer and the environment is completely poisoned.'

So our new urban farmer went home, lay in bed and thought about ethical food and a trip to Woodies the next day to buy some seeds—after all, at least Woodies is Irish owned. Armed with seeds of rocket, coriander, butterhead lettuce and a tomato plant already on the go, he was already feeling the glow of confidence that people like Alan Titchmarsh seem to emanate. It's all the gardening I'm doing, he tells friends when they comment on his healthy appearance. He didn't tell them that most of his seedlings didn't sprout and those that did peep out of his very expensive compost suffered a window box attack of super slugs on which he is now exacting revenging in the most unenvironmental way possible. He's even considered napalm, if it were available.

His second growing attempt worked out better, and after six weeks or so, he had five edible leaves from his butterhead and a sprinkling of rocket to add to the Cos lettuce in his regular shop from Superquinn. At least it's a start, he told himself while enviously watching the beautifully tended raised beds on *Gardeners' World*. I bet they don't weed that plot themselves, he thought, I bet they have loads of assistants and gardeners and make-up artists to put just the right amount of soil across Rachel de Thame's brow for that hardworking landgirl look.

WHAT DO YOU CALL THOSE ONES WITH THE *HUGE* HORNS?

However small and paltry his efforts, it still made him feel good to pick something green and edible from his own small property rather than from a supermarket shelf or that very expensive artisan food shop on Camden Street. He's been thinking about those damned chickens again. It's even extended to looking through Omlet catalogues for chic plastic coops that are a bit Philippe Starck if you look at them the right way. If he gets a bigger place next year, he will definitely put in one or two hens for eggs, and as someone told him plucking broilers is a right pain in the arse he might forget about chicken meat for the moment, though a wall-mounted neck puller sounds pretty impressive for the killing bit.

Our urban farmer has a long way to go, but he's feeling better about his life already. He likes the feeling of planting things, watching them grow and pottering about the garden like Diarmuid Gavin. His fantasy place is a twenty-acre smallholding in Meath. If he gets the promotion he's looking for, it might even happen in a year or two. Then he could even leave his job, grow organic veg commercially, keep some rare breed sheep, perhaps even get in some Highland cattle; aren't they the hairy ones with the huge horns?

However small a space we have, its possibilities as a food production unit have suddenly become very attractive. The huge rise in grocery prices since 2007 frightened a fair few people into growing their own small herb boxes and salad veg. Even if you only spend €3.50 weekly on a bag of salad leaves, this adds up to almost €200 a year. If you can plant a crop of rocket or mixed herbs for under a fiver, the economics of it makes a bit of sense. Many older Irish people are falling about laughing at the newfound novelty value placed on growing your own food. Until the 1970s, it wasn't unusual to see plenty of food growing in Dublin gardens, and in present-day city-centre 'artisan cottages', many back yards were home to the pig who proved his worth as both waste disposal unit and supplier of several months' worth of food.

But urban farming can be much more than just the small-scale stuff that fills a middle-class lunch table. Over in Cuba, when residents of Havana are not busy rolling cigars for foreign tourists, they manage to farm more than 35,500 hectares of small plots in the midst of the city's buildings, producing a huge variety of vegetables, spices, bananas, rice and other crops. This small-scale, mostly organic farming has created 350,000 jobs, twenty per cent of which are filled by women.[3] And it wasn't foodies who started it all; Cuba's urban farms grew out of the country's economic crisis in the 1990s. The cheap food and agricultural inputs that Cuba imported from the Soviet bloc kept the ship afloat, but after the fall of Communism, Cuba experienced a food crisis. Frankly, it was more like a hunger crisis, which Fidel Castro refers to euphemistically as 'The Special Period'. The food shortages triggered the urban farming phenomenon, which is still largely organic based. By 2006, farms in Havana and other urban areas produced 4.2 million tons of food, contributing to the establishment of a network of 1,270 points of sale of agricultural products and 932 agricultural markets.

The idea of growing food in urban spaces is hardly rocket science. It's just that it is very distant today from how we conceptualise food production. For decades we have seen food as something which is really only available to us in serried packaged rows on supermarket shelves, not freshly pulled out of the ground from a small plot beside the GAA grounds. Cuba's vast urban food projects are only surprising because they make us view food production in a new way: very little about growing food in urban areas is in any way technologically new or hugely challenging. For Irish urban and hobby farmers, the impetus to grow stuff is driven not by price, but mostly because we want more quality and taste. No supermarket carrot is going to taste as good as one pulled out of the ground in your own garden minutes before you eat it. Likewise, more interesting varieties of veg can be grown at home, as the seeds are not designed for mass production. It's not going to feed everybody, but for some people it's a way of getting the control and distribution of what they eat back into their own hands, and at a cost they can afford.

THRIFT IS THE NEW SPENDING

Alongside self-sufficiency and growing our own food, we are also being told that cutting our spending is the key to a more sustainable way of life. Magazines, newspapers and bookshops are full of advice on how to live with—horror of horrors—spending less. In the immediate aftermath of the global financial wobbles of autumn 2008, Delia Smith's book *Frugal Food* was suddenly back on the shelves. Sales of how-to guides to living with thrift rocketed, and in the weeks immediately following the worst of the banking collapses, Waterstone's bookshop saw a two hundred per cent increase in sales of two titles about keeping chickens. 'Carol Klein's *Grow Your Own Veg* sales doubled from what they were last year, which could be put down to people looking for long-term money-saving ideas in the current climate,' said Waterstone's non-fiction buyer, Alex Ingram.[4]

In Irish magazines and newspapers, articles abounded telling us to swap our clothes with friends, mend items rather than throwing them away and how-to guides to knitting jumpers from your Bichon Frise's moulted hair. Aside from the laughability factor, some of this very much brings to mind grandmothers and sucking eggs. While thrift became, ironically, something that represented the height of fashion, for anybody over the age of thirty and particularly for those who are older, thrift was pretty much the way most households were run in the Ireland of the past. This was back in the day before we all (apparently) hit the jackpot in the 1990s and turned from being a bunch of culchies into the Beverly Hillbillies. Yes, before we stuck our pickaxe into the ground and found oil, we had hand-me-down clothes, woollen tights with darned knees, home haircuts, and as for pet food, what the hell was that? Many of us who used to laugh at our parents' stories of walking seventeen miles to school through the fields every day were now, in the new age of recession, saying pretty outrageous things along similar lines. Perhaps thrift suits the Irish mentality. Perhaps being rich sat uncomfortably with us and inner penury is our psychological home address. Whatever the reason, living simply and thriftily was suddenly all the rage. But was it, really?

WE'RE BETTER OFF POOR

While many people breathed a sigh of relief at the economy quietening in 2008, few will rejoice at its collapse. When writers talk about the benefits of being in a non-boom era, what they really mean is that some people feel less pressure to be high achievers. In times of plenty, those who don't experience financial wealth feel left out of the party. But it's pretty hard to feel sympathy for those suffering with affluenza if you've never had it yourself. Downturns, on the other hand, often result in a quietening of the rhetoric of aspiration and a blurring of social divisions. The type of wealth we saw in boom-time Ireland can be divisive; it sends the rich into electric-gated private developments and their children into private schools, and it splits the world into the haves and the have nots. The Labour Party in the UK once denoted voters by where they shopped; an example being the Waitrose group, who worried about school league tables and climate change. If we were to apply the same model here, who would most of us be? In the present time, increasingly we belong to clans Lidl and Aldi, which says less about social class than it does about how the margins between rich and poor become blurred in times of less wealth.

But while cutting your spending and going back to basics was something many of us were forced to do by the downturn, to some onlookers, being thrifty is the mortal enemy of society. In times of recesssion, if people live thriftily and, crime of crimes, actually save money or reduce their spending, everybody suffers. This was the much-argued-about position of John Maynard Keynes, who pointed out that saving money leads to a reduction in consuming, which leads to a reduction in the production of goods, which results in unemployment, which harms society. Keynes was Mr Anti-Thrift in an era (1930s America) when thrift was not just something you read about in the lifestyle section of the *Sunday Tribune,* but more related to that hollow feeling at the centre of your belly.

Keynes considered thrift to cause social harm, a point that a lot of onlookers found unattractive, asserting that Keynes's book *The General Theory of Employment, Interest and Money* was also, at many points, disorganised, contradictory and confusing. 'How did it happen that a book so full of obscurities, contradictions, confusions, and misstatements was hailed as one of the great works of the Twentieth Century?' said Henry Hazlitt, not a great fan of Keynes.[5]

So what are we all to do? If we are good Irish citizens and alter our lifestyles to be more green and thrifty, grow our own food and buy fewer consumer goods, we could be all worse off in the long run, a phrase itself which used to send Keynes hopping mad—'in the *long run*,' he said, 'we are all dead'. Fair enough. But whatever your view of money or economic policy, the fact remains that in times of economic vulnerability, we are still being told to tighten our belts and stop buying so much stuff.

> The large growth in the Irish economy during the past 8 years has required a large increase in the consumption of raw materials and energy inputs, with consequential knock-on effects on the environment. The increase in economic activity is evidenced, for example, in the increase in the size of the labour force and in the level of fixed capital formation over the period. The increase in the use of natural resources can be gauged by the expansion in the area of land used for construction and the increase in both energy consumption and raw materials inputs. The ensuing increases in emissions of greenhouse gases further increases the pressure on the environment.[6]

So says the cso. This creates a dilemma here in Ireland and in other countries experiencing recession: buying stuff creates employment, but the rate at which we bought stuff, including food, over the past ten years helped create an environmental situation in which we are heading for the edge of the waterfall with no rudder on the boat.

Is being green, then, more important in the long run than keeping the economy going? The facts about our environmental future are pretty stark and, trendy lifestyles aside, unless we change our ways, the earth is still heating up at an uncomfortable rate, and not just for those wearing huge coats of white fur and lying around on icebergs. Ironically, Ireland's status as a green, rural isle is creating more issues for us than for other developed countries in terms of our carbon emissions and the speed at which we are contributing to global warming. Because of our unique position as a farming nation, we have lots of cattle. Four hundred and eighty four thousand of them. And lots of cattle produce lots of gas. While they happily roam Ireland's grasslands, cattle are actually giant fermentation devices, emitting greenhouse gases as they exhale and farting out fumes from the fermenting grass in their guts. Most harmful of all is methane, which is oozing out of those lovely soft muzzles at an enormous rate. The problem is well recognised, and research is being carried out in several institutions to examine how altering the diet of cattle, both beef and dairy animals, could limit the damaging effect of their greenhouse gas-emitting insides. Unfortunately, the reason Ireland is so good at producing beef and dairy products is our fantastic supply of grass. Giving cattle food supplements to limit carbon or introducing wide-scale changes to their diets is not feasible for most farmers.

GET RID OF THE COWS, LADS

So far, the solution seems to lie in limiting the number of cattle we produce, something that the UN wants to see happening all over the world. In 2008, its chairman of the Intergovernmental Panel on Climate Change called for a reduction in the global consumption of meat if we are to save ourselves from melting ice caps. But it's not just the emissions from agriculture that have to be cut. When Ireland signed up to the Kyoto Protocol, we undertook to reduce our carbon emissions by twenty per cent compared to our 1990 output before the year 2020. It doesn't look very likely. Industrial plants, electrical plants, dairy co-ops, etc. will all have to start paying for a proportion of their emissions within the EU's Emissions Trading Scheme from 2013, which will have unpleasant knock-on effects for consumers. While the downturn in the economy is reducing our harmful emissions from transport and manufacturing, the reduction is coming from a level that climbed alarmingly high in recent years. Agriculture accounts for a hefty twenty-seven per cent of our emissions per annum, but simply reducing cattle numbers does not solve the problem. Changing land from pasture to plant cereals or forestry does not significantly reduce emissions either, as tilling land releases large amounts of nitrous oxide, which is a particularly aggressive greenhouse gas with three hundred times the global warming effect of plain old CO_2.

Reducing the quantity of chemical fertilizers and introducing methane inhibitors are measures which are being investigated, but carbon taxes look to be the main approach that Ireland will take in tackling a reduction in our emissions. Some farmers say that allowing them to put some land into forestry to offset their emissions is a better approach, but so far under Kyoto, offsetting projects that capture carbon are less attractive than flat reductions in output. Minister John Gormley's warnings to farmers that the national herd will have to face cuts sent serious spikes of worry into the farming community. But at EU level, batting for the farmers' side on climate change strategy was Minister for Agriculture Brendan Smith, who was concerned that:

Given the relative size of the agricultural sector in Ireland's economy and the high proportion of greenhouse gas emissions coming from our largely animal-based production, the EU target posed particular difficulties for us, including the prospect of having to reduce our bovine herd.[7]

But there was, much to the relief of farmers, more to be said:

... the very significant reductions in agricultural activity that a 20% reduction in emissions would entail, would seriously affect the economic and social life of Ireland's rural areas, would involve negative consequences for the environment and biodiversity and would not be consistent with EU policy.[8]

WELL, WHAT ARE WE DOING HERE, FEEDING THE WORLD OR GREENING IT?

Here's the dilemma. While Ireland needs to be greener and to cut our harmful emissions, we still need to produce food for a world demand which is growing pretty fast. The demand for food suits the farmers, and Smith's arguments, and is one of the critical points at which environmental policy and food policy are not talking to each other or making a lot of sense. The reality is that food produced in Europe has a much lower carbon footprint than that produced in less happy environmental conditions in the emerging countries of the developing world. If we cut the number of beef animals, for example, on European grasslands, the slack will be taken up elsewhere, most likely in Brazil or Argentina, where the giant carbon-eating forests of the world are being cut down in order to expand land for beef production.

This is also evident in contradictory debates and government measures on growing biofuels and carbon-friendly crops. In 2007, the Irish government reduced its 2010 biofuels target from 5.75 per cent to four per cent, which was seen by some, especially those who grow the stuff, as a backward step in the promotion of growth, research and development on biofuels. Minister Eamon Ryan's reasoning was apparently that rising food prices across the world have been accelerated by land that is now farming crops to go into car engines rather than into people's mouths. But it's not as simple as all that. Rising food prices were massively helped by hedge fund speculators backing commodity prices, many of which, especially grain, fell dramatically in 2008. Yes, there is significantly more land now given over to biofuels than there was five or ten years ago, but some argue that in order to reduce carbon emissions in the long term, we have to make this change and make it fast.

The Irish Farmers' Organisation was puzzled, to say the least, by the government's backtrack on green policy.

> Considerable investment has been made by farmers and industry in developing an indigenous bio-fuels sector, but the move by Government to reduce the targets will jeopardise future investment in the industry. International experience has shown that the development of first generation bio-fuels is key to the successful transition to second and third generation bio-fuels at later stage. Blaming high food prices on the development of bio-fuels is simplistic.[9]

So it seems that even in Ireland we have still not resolved the issue in our heads. We want to reduce our carbon emissions and to encourage Pat the Plumber (if he still has any work) to swap his Hiace for a hybrid van, yet we are not providing farmers with enough economic incentives to switch over to producing the fuels that will make this change possible. Apparently, feeding the world is more important. But not if it's burgers they are looking for: having too many cattle is bad, and biofuels are starving us out of it. So where do we go from here?

GREEN ENERGY

Nationally we are making changes, and Irish individuals and families should pat themselves on the back for their efforts in recycling and changing to greener habits. Sourcing our household electricity from renewable sources (currently a very limited option in Ireland) and driving fuel-efficient vehicles are still very much tied to what manufacturers and power suppliers can make available to us. Wind and wave power in this country is still in its infancy. While companies such as Airtricity are making inroads in supplying renewable energy to domestic homes in Ireland, and more choice in the marketplace will be a good thing, the use of these sources of energy is, yet again, dependent on our attitude to erecting wind farms, wave power schemes, etc. What is clear is that the will to use renewable energies is here, and if energy companies give us the tools to live with less environmental impact, a lot of people in Ireland would take up these opportunities.

But as some scientists point out, finding a fantastic new problem-free source of renewable energy might not solve our doomsday environmental scenarios. Yes, it might halt our production of carbon and thus lessen the trajectory of disaster, but finding a new cheap and green energy source leads us again towards what's at the root of our problems living on earth: growth. Cheap energy means stuff can be made, produced, transported and sold cheaply. It provides employment; employment needs people and people need more food, water and other resources to sustain themselves. Thus we're back at the issue of using up more of the earth's resources than can be sustained, and ultimately starving ourselves out of existence. This is a pretty hair-raising scenario, and it probably won't happen, but often, trying to solve one environmental solution by coming up with another one to take its place doesn't work. It's the old problem of the blasted interconnectivity of everything. Changes that you make in your life certainly have an impact on the environment, changes people enact on a larger scale have a larger impact, and sometimes it's not possible to see what these outcomes are until they are upon us.

So what are we to do in Ireland? Is it worth making small changes if the larger ones heading our way are so insurmountable? Yes, is the answer. Having treated the environment as a giant factory to meet our needs for many decades, we failed to see the real, serious damage until it actually began to affect us and our perception of how viable it will be to actually continue to live on this planet. Now we have been shaken out of complacency, but the problems we are facing are, arguably, huge. But in the past we learned how acting collectively can bring about change, and have surprisingly positive results. Perhaps if we simply took stock of how we live, what we spend money on and what we eat, we might find ways to reduce both our environmental impact and our spending at the same time. In boom times, when the environment was given little thought, it turns out that not everyone was happy anyway. Being rich and throwing money around didn't come without its costs, and if we relegate this kind of thing to second place, we might get more enjoyment and personal pleasure from looking at the environment and food in a new, simpler light. Food shouldn't be that complicated; neither should living. If we make small changes, and enjoy the effects of making them, we can always make more.

PART TWO

Chapter 9 ∾

SUCKING DIESEL ON THE ROAD TO DUBLIN

Philip hangs out with protesting farmers

Huddled around a small radio, twenty farmers waited outside Kilmeen Creamery on the road to Clonakilty. Blowing into their cupped hands, they cocked an ear to the President of the Irish Farmers' Association, John Dillon, speaking to Pat Kenny.

'Farmers' incomes are in crisis. The average farmer's income last year was €15,000.'

'Sell your farms if things are that bad,' suggests Kenny helpfully.

'Who'd buy them?' retorts Dillon.

Tomato soup and ham sandwiches were laid out on a trestle table and nobody was shy about tucking in. Two hundred and forty-odd tractors had left Bantry in a 'tractorcade' earlier that morning. As they arrived, farmers stepped down from the cabs, massaging their backsides. 'It's the hips that suffer, 'specially with the suspension in that old yoke,' complained one, slamming the door of a rusting John Deere. It was a nippy January morning in 2003 and few, if any, of those ageing vehicles had any kind of cushioning in the driver's seat, and even less in the way of heating. Soup and sandwiches are not a luxury, but a necessity, particularly if you are going all the way to Dublin. Seven tractors had left Beare Island the previous day, possibly the furthest-flung corner of the country. Two would eventually make the symbolic journey all the way to Dublin. That morning, though, all seven, along with 230 others, left Bantry and started heading east. Bantry has a proud history as a staging post in farmers' protests. The Rickard Deasy-led 1966 protest, which shook the country and a complacent government, left from Bantry. But today they are also assembling in Castlebar in Mayo and Lifford in Donegal. Thousands are expected to take to the roads as reports come in from tractors amassing in Union Hall, Leap, Dunmanway and Skibbereen.

'I don't think we would have a big problem getting huge numbers and pure chaos on the roads if we wanted to,' an IFA official warned me. If I felt compelled to doubt him, the numbers of tractors that poured into Clonakilty dispelled the notion that this was the protest of a radicalised few.

'Listen to the Rustic Tiger roar!' he yelled to me with a wink, gunning over five hundred horsepower of Massey Ferguson.

Margaret Brennan was out on the road with two of her three children, counting the tractors passing. After 122, she left the counting to the children and told me about how four farms in the parish had sold up in the last two years, and many more farmers had gone out to find work elsewhere: 'Driving lorries, doing brickwork and the like.' With the flight from the land, her husband's contract work on farms had dried up and they were now trying to support a family of five on weekly benefit payments of €248. 'I hope the government listens to the farmers now. We just don't be heard so I hope they listen to those who are still farming.' But as she pointed around her to the unlit houses along the road, it was clear which way the trend was moving. 'I do see a lot of people who have sold up. There's one just up there. There's another just one mile down the road that had to sell up and go to town to find work. They haven't the money, like, and nobody in Dublin understands that.'

There was an air of spectacle on the way to Clonakilty. Mothers and children dotted the crossroads and counted the tractors passing by. Tradesmen and lorry drivers pulled over and waved a salute rather than try to race ahead of the convoy. By the time 350 tractors reached Agriculture Minister Joe Walsh's hometown of Clonakilty, the small knots of ones and twos had expanded to dozens lining the street. No horn was spared as they passed the minister's house on Emmet Square, and the crowd loved it. 'Are ye hearing that now, Joe?' roared one supporter. He would have to be listening very hard to hear: the minister was 'attending meetings all day in Dublin', according to his spokesperson.

The IFA leadership had repeatedly declined to say what, if any, specific goals they had. The tractorcade set off on a Monday, and bizarrely, John Dillon said he would only articulate their claims on Thursday as they neared Dublin. 'How very French; demonstrate first and work out what you want second,' sniffed one of my journalist colleagues.

Listening to Dillon as I gave him a lift into Bandon, it was hard not to form the impression that he was playing a far more savvy game than anybody had reckoned on. He knew that farmers and the agricultural sector had been fighting a losing public relations battle for the last ten years. It had started when the Beef Tribunal had revealed the rotten and corrupt heart of the cattle barons' empires. Greed, environmentally damaging practices and a growing awareness that paying farmers guaranteed prices was not only unjust but unsustainable had further tarnished their reputation, and all this before the killer blow: variant CJD. The farmers' and food processors' pursuit of profit at whatever cost had turned cattle into cannibalistic carnivores. There wasn't enough protein in grass to encourage growth quickly enough, so why not feed the protein-rich meat and bones of dead cattle, the parts that humans won't touch, back to their brothers and sisters? Watching BSE-diseased animals stagger around farmyards had been enough of a Frankenstein nightmare for consumers, but now the disease had made the leap to humans. Irritability and loss of concentration in an unfortunate few had been followed by failing sight and motor function, leading to a slow and inevitable death, and all so that some could add the halfpence to the pound.

John Dillon knew that after the food shocks of the nineties, the farmer's stock was low with the Irish public. But he had also seen how the same public had rallied around the farming community when foot and mouth disease arrived. Without a second thought, the entire country had slipped into national emergency mode because that was what was required of them. Small sacrifices were made by all in support of the farmers. Dillon was right in so far as there were very few households in any town or city where there wasn't a farmer in the extended family. Many suburbanites were barely even a generation away from toiling on the land. The Irish people were connected to farms and the plight of farmers even if they didn't know it. Dillon wanted to reconnect that constituency with his. If the Irish people could only hear about the lives of quiet desperation that were being led in small farms up and down the country, they would set aside their negative opinions.

Over the course of that first week of January, those stories of genteel and outright poverty were heard. For a change, the stories were not of headage payments and the value of REPS schemes and other unintelligible technical agri-jargon. I listened to the embarrassment of a two hundred-acre tillage farmer who for the last two years had to make a choice as to which one of his three school-going children would get a pair of runners that year. The farmer's wife who never threw out a teabag before it had made at least four pots of tea. When did you last have a foreign holiday? 'Never had one. Ever.'

Only the most foolhardy of gamblers would choose to get into farming. If you take a punt on the stock market you can spread your investment about in case one stock underperforms. If you open a clothes shop you buy in a number of lines for fear one may prove unpopular. When you farm you get just one roll of the dice. Weeks and months of careful and loving husbandry can turn to a pulpy, worthless mess while it is still in the ground after just a few days' bad weather, and you don't get another shot at it until the following year. In Fermoy, one man told me how when that happened you just closed the curtains and got back into bed. 'And you know that the only thing in the post that year is going to be a tax demand from the Revenue.'

The farmers deserved much of their bad press. They needed to hear the justifiable criticisms that were being levelled against them. But they had a human side to their story that hadn't been clearly articulated before. As the week wore on and the tractors got closer to Dublin, it was apparent that John Dillon had succeeded in stimulating a genuine national debate. If there was an argument to be won, they might be winning it. A debate had begun, and for all their perceived and actual sins the farmers were getting a sympathetic hearing.

The tone of newspaper articles and headlines changed. They still mentioned traffic disruption, but it was buried deeper inside each piece. Complaints to talk radio shows, which had been shrill at the outset, became muted as town and country folk grappled with the issues at hand.

I can't just sell my farm in the same way a bankrupt business can sell its premises. You can turn that business into any kind of other business. A chip shop could become a solicitor's office. An unprofitable farm remains an unprofitable farm unless it gets rezoned.

Yes, the farmers were still a terrible lot of whingers, but the penny was dropping as to what their whingeing was about. Some town and city dwellers realised that their own lifestyles were enabled by an enormously costly, taxpayer-funded infrastructure, with a choice of schools, hospitals and entertainment their country cousins could only look at enviously. They were not so far removed from the farm that the farmers' arguments didn't make some sense. The gap in understanding between town and country began to seem a little bit narrower. The entire country was talking about the urban/rural divide, about life west of the Shannon, food safety, hidden poverty, about what Ireland needed to do to properly earn the title 'The Food Island'. Or so it seemed to this reporter at any rate. Perhaps I, and others like me, overestimated the impact of the debate. It wouldn't be the first time a journalist got swept up in the emotion of the moment and completely misjudged what the real impact of an event or movement actually was. But as we stopped in Fermoy, Mitchelstown, Urlingford and countless other towns on the road to Dublin, there appeared to be a genuine consensus of opinion. Not about what should be done, but at least that this was too important a debate to ignore.

The tractorcade moved inexorably on towards Dublin. Five days after seven tractors had sucked diesel up the Beara peninsula, thousands had assembled in Wexford, Kildare and Meath, readying themselves for the final push on the capital. As dusk fell on the eve of the Merrion Square rally, I met one of John Dillon's more jaundiced lieutenants in a farmyard in Ashbourne. Initially unenthusiastic about the whole endeavour, like many others he had been won over by the warm reception the protest had got thus far. We joked about the legions of tractors in a ring around the M50 being like Caesar's army poised to cross the Rubicon. Did he think that they would get much support from the people of Dublin? 'Wait till you see now tomorrow, how many Dubs are actually from the country. They'll be lining the streets three deep,' was his buoyant assessment.

I could see his point. In 1971, almost half of the country lived in a small village or out in the country. By the time the tractorcade arrived on the outskirts of Dublin, thirty-two years later, that figure was closer to a third of the population. Logic suggested that many who had fled the land would now be found inside the ring road motorway. Countless adopted Dubs had to have aunts, uncles, brothers and sisters still working a farm. They would have memories of dragging the milk urns to the stand at the bottom of the drive, of a newborn calf rising shakily to its feet, of the agony of a ruined harvest. Surely they wouldn't be so far separated from their rural roots that they would have forgotten the 24/7 nature of the vocation that is farming. 'Farming is a part of what we are. Irish people all come from the land at some point, no matter where they live now,' was how George O'Hagan, a Donegal sheep farmer, put it.

A representative sample of three hundred tractors was chosen to complete the final leg of the journey. There was much whooping and cries of 'Up the Banner' and the like amidst unnecessary gunning of engines as they saddled up and moved out. Slightly incongruously, John Dillon sat astride 575 horsepower of tractor in a suit and neat blue overcoat as they passed under the motorway and into Glasnevin. If they were expecting a tumultuous welcome—and they were—they were to be disappointed. The silence was so resounding it was as if metaphorical tumbleweed was blowing across the Navan road. Past the cemetery and Botanic Gardens and on into the city centre, few paid much, if any, attention. A flotilla of tractors! There's something you don't see every day, let alone in the middle of Dublin. Yet jaded jackeens barely looked over their shoulders as they went on about the business of getting to work.

Parked four deep around Merrion Square, the farmers set about congratulating themselves heartily on how well marshalled and disciplined they were, but they were talking to themselves. In contrast to every town along the route, where cheering locals lined the streets, Dubliners were indifferent. 'Why are they blocking up the street with them … things?' asked one office worker, unable to put her finger on the word 'tractor'. One particularly vexed businessman grabbed my microphone to register his complaint.

'What right do they have to be here?'

'The same right as anybody else to protest, I suppose. Do you support them?'

'No I don't!' he bellowed. 'I can't find a fucking parking space.'

John Dillon made a speech, the same one he had made every day that week. The farmers revved their engines. Their former leader, Tom Parlon, at that point a junior minister in government, made a brave appearance. More revving of engines. He had a good-natured row about the price of rolled oats and disappeared. Yet more engine revving. And then that was it. They limped off out of the city to no more public attention than they had got coming in. The protest appeared to have come so far in public perception in just one week, yet it hadn't got anywhere at all. In cold political terms, it hardly mattered how much support it got on the streets of the capital, but a rousing reception would have been the crowning glory to a week of positive press. However, the reception had been so underwhelming that nobody could bring themselves to speak of it. The urban sophisticates had greeted the farmers with apathy. Stunned, they drove their tractors onto low-loaders to drag them home again (no sense suffering that hip-crushing journey twice).

Outwardly they had done exactly what they set out to do: flex their muscles and bring an ill-defined set of complaints to the doors of Leinster House. That was all they did, though. For a moment some had thought that the gap between country and city would be bridged. If anything, that week revealed exactly how far apart the two had been driven. It was as if the whole endeavour had been the metaphorical tree falling in the proverbial distant forest, unwatched and unnoticed, and therefore utterly irrelevant to the lives of anybody in the capital.

That Dubliners didn't care about an incomes protest mounted on behalf of barely four per cent of the labour force might not be surprising. But in turning their backs on the protest, what were Dubliners saying about their attitude towards food? Wasn't food a problem that had been solved for the Irish? No need, therefore, to care about what the farmers had to say. Was it all completely irrelevant to them because they knew they could get their beef from Brazil and their all-year-round asparagus from Israel? Were they saying, in essence, 'We don't really give two hoots about anywhere west of Naas'? 'We have left the damp, depressing reality of country life generations behind us for a bright new future of Aer Lingus weekend city breaks and hot and cold running Latvian housekeepers'? You can be sure that when Michael O'Leary makes a pronouncement on aviation fuel prices, the middle classes prick up their ears. That's because there's still only one way off the island. But when farmers grumble about one hundred per cent increases in fertilizer costs, the silence is deafening.

The entire history of human development has been about food, and about securing cheap and plentiful supplies of the stuff. Wars have been fought over it, empires have collapsed for the want of it. It is only relatively recently in this country that we have been freed up from the daily grind of expending enormous amounts of energy securing our next meal. Little wonder that the Irish had such a lust for land. Land is the growth medium for food and producing food was the most important thing you did every day. Now, for far too many of us, food is solely the means to get rid of that annoying sensation in your belly. We don't have to worry about where we are going to get it from, it's just there when we want it. We put our hands in our pockets and the grumble in our tummy disappears. That freedom from the search for food has brought about an enormous social revolution.

If the history of mankind's progress out of the trees to just popping around the corner for a panini were a twenty-four-hour clock, we achieved food security at about a minute to midnight. After the Second World War, we threw everything into never going hungry again. We achieved that and it has given many of us the time every day to do more interesting things than hunt and gather. It has profoundly reshaped society, probably for the better. The majority of the most stunning advances in Irish society have been made in the last fifty years. Many of us are now free to pursue ambitions and careers that would otherwise have been unfulfilled. Women in particular were freed from lives of appalling domestic drudgery as the food industry became ever more innovative. Convenience was its watchword, and what's not to love about convenience?

The flipside of this coin is a troubled one. Mirroring our shift to becoming a post-agricultural society is our more tricky relationship with food. It was only to be expected that as more of us moved away from being involved in food production, the less we would care about how it was done.

But now we are two clear generations away from knowing when a food is seasonal. It is probably a long time since a mother stood at a meat counter and told her daughter what mutton neck tastes like and how you can prepare it so that it is even more tender and flavoursome than lamb. We might not have needed much of that bathwater in our lives, but we shouldn't have chucked the baby out too. Now, for the vast majority of us, there is no other daily activity we do more frequently and into which we put less thought than eating. True, that is our prerogative, but where the economic majority go, the rest will invariably be forced to follow as food becomes more convenient and less healthy. Ireland's food-related illness is a bill that must be shouldered by all and runs into the billions. For the rest, food is a source of anxiety, their interaction with it massively more complicated than when we had to struggle just to get enough. They stress about how it changes body shape, shortens lives, potentially poisons, is a tool of economic exploitation, or is just too expensive. While food is now plentiful, many of the methods of producing greater quantities of it have done neither our health nor the environment any favours. The cathedrals of plenty and convenience where we do our weekly shop have also had the most enormous effects on the shape of our lives. A lot of time has been spent looking at how television, the internet, global travel or easy credit have changed Ireland and the Irish. There are few, if any, studies that have examined food as a factor in social change. This second half of the book is the next chapter in the story of our flight from the land. Now that we don't produce food ourselves, who does, how do they do it and how is it changing Ireland?

Chapter 10 ✿

| EVERY LITTLE HELPS

What happens when Tesco comes to town

Time was when there was a lucrative sideline in supermarket openings for Ireland's politicians and B-list celebrities. By midday on a Friday there would be a traffic jam of presenters leaving the RTÉ campus in Donnybrook for the provinces. They took to the roads on the promise of a fat cheque in exchange for a few cheery words about how they never knew happiness until they had visited their local version of this particular supermarket chain.

The rent-a-crowd of the manager's friends, bulked out with Ukrainian staff members dragged in on their day off, all remark on how the presenter's dye job doesn't work as well in the flesh as on television. Stuck to the celeb's elbow is the local government party backbencher, the one whose name nobody ever seems to know. His dilemma is whether he should leave and try to squeeze in attendance at another funeral before the ribbon-cutting ceremony, or wait until the snapper from the local paper turns up. You may not have been to one of these openings, but you've certainly seen the photo somewhere between pages four and seven of your local paper. In addition to the three hundred words about exciting new retail opportunities for the town, there'll be a picture of the supermarket manager overenthusiastically clutching the waist of the kids' TV presenter. The TD stands behind them wearing the same rictus grin he used overleaf on page eight for the dedication of the new fire tender. Around them, the semicircle of balloon-waving Slavic shelf stackers and check-out staff are wondering if the photo op will be over before they're due to start their split shift in the pub.

It was a useful symbiotic relationship for all concerned. Publicity-hungry minor celeb currently 'between gigs' gets exposure; their presence guarantees that the photographer actually gets out of bed; the politician's business creation credentials are polished; and the supermarket gets to make a news event out of something that would otherwise be less read than the minutes of the last local authority council meeting.

However, all has changed utterly. In the last few years, hundreds of stores have been opened without the vital assistance of a presenter/model/ actress/whatever. Could it be that the demand for minor celebrity attendance at supermarket and convenience store openings has grown beyond what Ireland's small pool of media big shots can supply? In Nenagh, Tullow, Ringsend, Kilrush, Ballylough and Inchicore, Tesco has had to open new stores without the ribbon-cutting skills of any recognisable public figure. In fact, they have gone as far as creating their own VIPs for a day from local in-store competition winners. Congratulations to eighty-four-year-old Maureen Synan from Cooraclare, who officiated at the opening of Tesco Kilrush in February 2008. At store openings in Camden Street, Talbot Street and Parnell Street in Dublin, there wasn't so much as one goose-pimpled, perma-tanned Irish superstar to bring any homespun glitz to proceedings.[1]

Worryingly for Ireland's household names, this erosion of their public duties is only set to get worse. As the economy contracts, one of the few companies boasting about expanding is Tesco. In terms of square feet of shop floor, in 2008 they opened five football pitches worth of retail space and almost six the year before. At a time when every other sector of the economy was battening down the hatches, Tesco boss Terry Leahy was saying that nothing was going to curtail the 'rapid expansion in Ireland' of the supermarket chain.[2]

There is, of course, no shortage of Irish wannabes who would jump at the prospect of officiating at a supermarket opening: this remains one of our healthiest growth industries, with a steady stream of new entrants to the marketplace. The likely ulterior motive behind denying celebs a handy nixer probably has much more to do with how Tesco wants to be perceived. This is very important for a company that has weathered a storm of negative publicity about its detrimental impact on local retail and its poor treatment of suppliers. As Tesco opens up stores in Ballinrobe, Bailieborough and Bettystown, the brand is eager to underline what it gives back to those communities. Ribbon-cutting duties have passed to the managers of local resource centres or representatives from charities like St Vincent de Paul or the Samaritans. This is no mere token gesture, though, as the resource centre will also get a cheque for €1,000 from the local Tesco manager.

Before the fizz started to leak out of the global boom, Tesco's turnover in

2007 was about €60 billion, with €3 billion of that being generated in Ireland.[3] With each new shop opening comes the promise of jobs for the area. Ninety-four in Tullow, County Carlow. One hundred and thirty in Nenagh, County Tipperary. And so on, adding to Tesco's fourteen thousand-strong workforce in the Republic alone.

While these low-wage jobs are no doubt welcomed by those who get them, unhappily we are not too good at recording how many jobs are lost when Tesco comes to town. If it's anything like the UK's experience, and there's little reason to assume it wouldn't be, it is disastrous. Over there, the National Retail Planning Forum studied what had happened in the areas around almost one hundred new superstores. They noted a five per cent drop in food retail-related jobs in a fifteen-kilometre radius. The stores they studied were responsible for the loss of over twenty-five thousand jobs. In the Forum's own words, 'if the superstores had not opened, employment would have risen'.[4] Possibly the most embarrassing aspect of this report was that the National Retail Planning Forum was not some bunch of academic lefties lobbying for a farmers' market in every town centre. Their research was financed by Tesco, Marks & Spencer, and Sainsbury.

THE TOWN THAT TURNED ITSELF INSIDE OUT

Seven o'clock on a June morning and there isn't a single soul anywhere in Nenagh's town centre. There might just be a distant hum from the bypass as early risers head towards Limerick, but otherwise the Tipperary market town is completely at peace. Three doors down from the junction of Kenyon Street and Pearse Street, the words 'Bar and Grocer' were set in relief into the stonework above the shop front in perhaps the forties or fifties. The person who did that probably never imagined the contradictory brash red Vodafone signage beneath. Directly opposite, the shutters are pulled down on the old Dunnes Drapery and they won't be lifted when everything else opens. Flann Lowry is the first to pull up the shutters of his newsagent's and he gets straight to work, slicing open the bundles of newspapers with a Stanley knife. But it will be a while yet before he sells his first paper this morning. The crows gathered on the telephone wire look down on Hanlon's and Quirke's butchers, but there's no activity here to get excited about yet. Nenagh could not be more at peace with itself at this hour.

Ten past seven, a little bit out of the town on the Borrisokane road, is an exercise in stark contrasts. Tesco's doors are wide open for business, as they have been all night. Twenty-foot-high banners in each corner of the car park proudly announce 'Open 24 Hours'. There are already thirty cars parked in the 250 available spaces, though the majority of those probably belong to staff. Inside, one wonders might the range of choices from the twenty-one food aisles and twenty-three other avenues of clothes and household goods not be a bit dizzying for this hour of the morning. A thirty-seven-inch Grundig flat-screen TV, or the cartoons it is showing, draws the attention of a six-year-old girl who has broken free from her mother. One aisle across, the mother runs her hand thoughtfully along the Acer laptop computer. If Dad were here he would be able to pick himself up a machine-washable men's suit in plain black or pinstripes for just €38.[5] By eight o'clock, three sets of parents with uniformed kids in tow have all followed the same path to the pre-packaged sandwiches, swinging by the fizzy drinks display, fruit and veg for an apple, then a packet of crisps and on to school. Valerie has also picked up a few bits of her weekend shopping at the same time. 'I used to shop in town all the time, but I've just got into the habit of coming here.'

Back on Kenyon Street it's twenty past ten before anybody walks into either Hanlon's or Quirke's butchers. By half past ten the fruit and veg shop beside them still hasn't seen its first customer. In NewsFayre, Flann Lowry is busy selling papers and the odd bottle of water, but footfall is not what it used to be. 'Tesco is doing all the same business we are. The only thing that has saved us to date is that we are on the main street. If we were anywhere else we would have shut up shop a long time ago.' Before Tesco arrived in the summer of 2006, Nenagh's retailers consoled themselves with the thought that the supermarket would only be selling food. Few contemplated that it would also be a one-stop-shop for newspapers, books, CDs, gardening supplies, cosmetics and hardware. It's very hard to resist that kind of convenience under one roof. Flann and Mary Lowry's greetings card shop is still doing good business, but of their newsagent's Flann says, 'There's no way we'll still be here in ten years' time.'

Particularly when the county council seems to have done its unwitting best to drive Nenagh's consumers away from the existing traders and out the Borrisokane road to Tesco. One planning decision after another has turned Nenagh completely inside out. The focus of the majority of the town's commercial activity is now up to a kilometre away from the town centre. While there are several brownfield sites in the town that are crying out for development, the council has allowed Tesco, Dunnes Stores, Lidl and SuperValu to develop on greenfield locations on the outskirts. Between them they offer well over one thousand free car parking spaces, where in the town centre the cash-strapped local authority has introduced pay and display parking. Hoping that some of this parking revenue would be ring-fenced for a multi-storey car park, some leading traders identified and negotiated the purchase of a site on the council's behalf. The council took the site but didn't build on it; they also took the parking metre revenues and built themselves new offices on—wait for it—a new greenfield site opposite Dunnes Stores. This latest in a long list of dumb choices delivered the sharpest slap in the face to Nenagh's existing business community. Moving the council and HSE offices took a couple of hundred people out of the town who used to shop on foot every evening. Now they go to Dunnes. 'Since then you're waiting for a five o'clock rush that just doesn't come,' says delicatessen owner Peter Ward.

Footfall, say the traders, has dropped by fifty per cent in the few years since Tesco arrived. The choice really makes itself for Nenagh's shoppers, faced with free parking right at the door of a supermarket shop where they can get everything they need versus feeding the metre a fiddly forty cents every hour for hard-to-find on-street parking. Few of the town's businesspeople will talk freely in a way that runs their ventures down, but the difficulties they are experiencing are obvious. The farmers' market, which had been building its Saturday morning trade quite successfully, will probably be the next casualty. Already it is fragmenting as individual stallholders are throwing in the towel. All the supermarkets source their meat centrally and ship it in from outside the county (and frequently outside the country) every day. Nenagh's five traditional butchers all source their produce from local abattoirs and farmers. They are all aggressively optimistic about the future, but adding their five premises to the four in the supermarkets, one wonders whether Nenagh might not be a bit over-butchered.

There is a world of social difference between the two shopping experiences on offer in Nenagh. Whether at seven in the morning or the evening, Tesco's shoppers move silently up and down the aisles. Perhaps stopping for a chat while choosing between four different brands of sea salt is too great a feat of multi-tasking for most. In Peter Ward's delicatessen or Flann Lowry's newsagent's, though, there isn't a single customer not greeted by their first name. Each has a titbit of news to impart or receive.

In fact, in one of Nenagh's oldest institutions, the exchange of such vital information is all that seems to go on. The country market was started forty-two years ago by the ICA. For all but one week of the year, up to a dozen farmers' wives take over the Catholic Church hall every Friday morning. Luring people off the streets with damson jams, home-made butter and apple pies that you can 'pass off as your own', they say they have managed to weather the competitive storm to which Saturday's farmers' market looks like succumbing. Una Caulfield is adamant that nothing has changed about the market since the multiples came to town. They are as busy as they have ever been. Looking around at the trestle tables of pies and jam jars, there seems something at odds with the market's 'Crafts and Foods' billing. Where are all the crafts? 'Oh, we don't bother with the gifts and craft end of things any more. Not since Tesco came. They do that sort of thing much better.'

The same tensions at play in Nenagh are to be found in twenty-five other counties and countless small towns. In Mallow, the Chamber of Commerce warned that the Cork town could become 'little more than a commuter belt around a dying town centre'. Another twenty-four-hour Tesco in Clonmel is wreaking havoc with some of that town's more established traders. Letterkenny has also turned its geography inside out to accommodate the supermarkets. The same mistake is being replicated again and again as local authorities appear happy to absorb the supermarket business model into their county development plans. 'You want to build big? Grand. You want to build a mile away from the town centre? Okey dokey!' This allows the supermarkets to compete on their terms on their turf, not so much an uneven playing pitch as giving them a pitch of their own. The Competition Authority's response to food price inflation is naturally more competition. Unfortunately, they don't mean increased competition between the multiples and traditional independent retailers. The kind of competition they want to see is between the supermarkets and the hypermarkets, which can be achieved by securing the entry of more multiples into the Irish market. And they want the planners to increase the amount of floor space on offer.[6] Make way for Carrefour, Sainsbury and Wal-Mart.

What unfolded in Britain as the supermarkets expanded was a lesson for us to learn. Did we not suspect that the estimated two thousand independent shops that went out of business every year was something that might happen here? Or was that just a price we were happy to pay for the sake of convenience?

RESISTANCE IS FUTILE?

If the combined floor space of Ireland's major supermarket chains was two lanes and a hard shoulder of the Dublin–Cork road, it would run from the Red Cow roundabout to Abbeyleix. Growing by up to five kilometres per year, it will have reached Cork's county border in eighteen years' time. If Tesco,

Dunnes, Superquinn, Aldi, Lidl and Marks & Spencer were national stadiums, we would have 204 Croke Parks. Over a square kilometre of Ireland is now under the warehouse roof of one of the big chains.[7] And that doesn't include any of Spar's four hundred-plus franchises or the SuperValus and Centras.

Tesco, Dunnes, Superquinn et al. will claim, and not unreasonably, that there is a good reason why they have achieved pre-eminence. Quite simply, they are able to offer better value, more choice and the utmost convenience to the shopper. Food snobs and traditionalists who complain about the way they do business are missing the point. If their rivals had been able to offer what the supermarkets can, they would still be in business, wouldn't they? Bemoaning the spread of supermarkets is like blaming the fox for breaking into the chicken coop and not confining himself to rummaging in bins. It is their nature to expand. If they don't, they won't survive. Businesses of that size can't rely on one or two bumper years to tide them through the lean times. Their life blood is growth. They need to grow every year. When Billy Busybody from An Taisce gripes about ugly warehouses on the outskirts of every Irish town, he might as well be admonishing the fox for being a fox. Rather than criticise them for being what they are, we should ask ourselves, why did we leave the door to the chicken coop open?

Supermarkets are big not only because they want to be, but also because we allow them to be. When Tesco said they wanted to build on the outskirts of Nenagh, why were they not pointed towards the brownfield locations in the town? Surely Tesco wanted the opportunity to do business in Nenagh just as much as the people of Nenagh wanted to do business with Tesco. Why do business on Tesco's terms? Why wasn't the welcome mat rolled out, but only as far as one of these sites? By all means make them feel at home, but alongside the town's existing traders, which is pretty much how H. Williams or Quinnsworth did business for years, operating cheek by jowl with independent traders. This would offer the consumer more genuine competition and choice. But instead, in countless towns around the country the planners have accepted the supermarkets' business model as what is socially desirable, so now we shop in warehouses large enough to comfortably accommodate two elephant stampedes in opposite directions. Faced with the choice between free parking close to the shop door or inconvenient and expensive town centre parking, shoppers have voted with their family saloon car and not their feet. Accepting this model as best practice for development has handed the supermarkets an enormous competitive advantage. Do the planners think that if Tesco isn't allowed to build as big as they want, where they want, they won't come?

PHILIP GOES SHOPPING FOR SUNFLOWER SEEDS

'I should have you here all the time,' says Fergus Dunne ruefully, looking at the inside of his till. In the last twenty minutes about as many people walked into his shop. They spent an average of €5 each. But that's not how things have been going. There should be plenty to pull in Tullamore's passing trade. Fergus's organics store, Wild Harvest, is in a prime location on the corner of the Square. Cleverly arranged displays on the footpath present everything from chilled fruit smoothies to organic fruit, veg and cooked meats at prices consistently lower than the supermarkets. In a supposedly health-conscious era, you'd think Fergus would be doing a ripping trade. You'd think.

Fergus got into the food business after a year travelling in South-East Asia. He became captivated by Malaysian people's relationship with their food, and how they would buy fresh produce every day and prepare a meal straight away. He returned home and took over the premises from a beautician in June 2005. He either didn't notice, or through blind faith in his vision chose to ignore, that Tesco had opened up a petrol forecourt alongside their Tullamore supermarket just two months earlier. In retrospect he does remember the beautician cautioning him about the decreasing number of people shopping on the main street, but he trusted in the good people of his hometown and pressed blithely on. A few years on, he says, 'If you get a wage you are doing well.'

His dilemma has been that Tesco are brilliant and utterly ruthless in identifying what is selling well in other stores and then taking the product on in their own store. Pointing out a display of Linwood's flax and sunflower seeds, Fergus says that they were a great item for him. But then Tesco got them in and, even though they sell them dearer than he does, nobody gets them from him any longer. 'They don't have to sell them cheaper; it's just the convenience factor.' With the kind of acceptance born of inevitability, Fergus has a pretty good idea what his fate is going to be. At a time when it seems to him that there are more cookery programmes than news programmes on television, 'there's an awful lot of old talk about food. But when it comes to the crunch people are not concerned enough to walk the walk.'

When he goes, who else is going to convince themselves that setting up in competition to Tesco in Tullamore is a clever business plan? In short, nobody will. Food provision is becoming concentrated into an ever smaller number of hands. Tesco, SuperValu and Dunnes Stores basically control the supermarket sector between them, with Superquinn really only mopping up in Leinster. The vast majority of the 2,500 franchised convenience stores get their goods from the same two wholesalers, Musgraves and BWG. BWG are intent on getting bigger and bigger: not content with supplying all the country's Spar shops, in 2008 they bought up one of the last few independent wholesalers, Mangan's, who supplied the Mace and Vivo franchises. BWG had already signalled its intentions towards the Londis stores in 2004 with a buyout bid that was rejected, and now they are focusing their competitive fire on Musgrave's, targeting their Centra and SuperValu stores.

Half of the groceries bought each year come from one of three shops: Tesco, Dunnes or SuperValu.[8] Before long, the majority of us will be getting everything we eat from one of just a half dozen suppliers. Would motorists be happy if there were only six car brands to choose from? Would we be living in a democratic society if there were only six sources of news and current affairs to choose from? This is about our food sovereignty. Can we be happy to hand that kind of control over what we eat and where it comes from to such a small number of companies? While we like to think of ourselves as free economic agents with the power to influence how the supermarkets behave, the reality is that they control how we behave. They lure us in the door with promises of better value that most of us are too frazzled and busy to scrutinise properly. If we did, their cynicism would be revealed. In 2008, Tesco responded to falling sales by slashing prices on ten thousand items. A National Consumer Agency price survey revealed that they had secretly jacked up prices in the weeks before they secured enormous publicity with the cuts announcement. In May 2009, Tesco reacted even more ruthlessly to dwindling market share. They trumpeted cuts of an average twenty-two per cent on thousands of items to lure recession-weary shoppers back across the border from Northern Ireland. Little did the prodigal consumers returning to Tesco's aisles from Asda and Sainsbury know that the savings were being made at the expense of Irish jobs. Rather than feel the pain of the price cuts themselves, Tesco passed them on to Irish food producers and farmers. Already struggling through the downturn, these producers cut their businesses to the bone to stay alive. 'It wasn't a negotiation, it was a bloodbath,' said one who was forced to slash his price by fifteen per cent or lose Tesco's business. Consumers who thought they were doing the patriotic thing by spending their money south of the border were actually contributing to the loss of jobs in the Republic.

In Tullamore, Fergus Dunne knows that he is living on borrowed time.

The butcher around the corner from him is closing after thirty years in business. Neither will be replaced by others willing to take a crack at a venture that is proving to be a sure-fire flop. Whatever about his own fortunes, Fergus worries about Tullamore ten years from now because his suspicion, albeit coloured by bitter experience, is that, 'once they've killed the town centre, they'll up their prices'.

Like the Borg, the all-conquering alien race in *Star Trek*, resistance in the face of the supermarkets is futile. Since 2001, the number of supermarkets has doubled. While that was happening, in the five boom years before recession half of the country's independent retailers were driven out of business. There are now fewer than 3,500 of these kinds of businesses left.[9] What has happened here is, if anything, worse than our nearest neighbour has experienced. During the 1990s, as Britain's number of supermarkets doubled, it lost a third of its smaller shops (those with a turnover of less than £100,000 sterling).[10] The worry is that as our smaller shops shut up, a tipping point will be reached where wholesalers no longer find it profitable to deal with independent retailers at all. In Britain, a cross-party parliamentary report found that by 2015, any independent grocer or convenience store was 'unlikely to survive'.[11] If we keep losing shops at the rate of about five hundred a year, as we are now, we will be only twelve months behind Britain. In a very short time you won't be doing any kind of shopping that isn't in a supermarket or franchised convenience store such as Spar, Mace or Centra.

Some planners in more innovative local authorities have started using existing planning laws to try to protect and enhance diversity. Enniscorthy is one of the better examples of how it can be done. Either side of the River Slaney, the streets buzz with life. The shop fronts are a healthy mix of butchers, hardware stores, greengrocers and chemists. Because the town is set in a steep valley, it's a calf muscle-busting climb to get from one shop to another, but that is how many people do their business. Up and down, up and down with trolleys, canvas bags and baskets slung over the handlebars of pushbikes. If it's convenience you are looking for, there are two large supermarkets with big car parks. Crucially, though, they are right in the town centre and not out on the periphery. The Dunnes Stores is tacked onto a creatively converted mill and the SuperValu on the river sits cheek by jowl with the town's other businesses. The opportunity for genuine competition between small retailer and supermarket exists, and both appear to be thriving.

Traders know what they have and how lucky they are not to be slipping towards becoming another ghost town. Perhaps it is Enniscorthy's hilly topography, rather than genuine foresight, that has corralled the shops and supermarkets together into the town centre. Whichever, the town is all the better for it and it points the way towards what might yet be made a planning requirement for any of the multiples. Build by all means, but only on brownfield sites. Force the supermarkets to become instruments of urban regeneration rather than agents of a town's slow decline.

Enniscorthy is a positive model of how independent traders and the multiples can exist together. However, after we left Wexford, a telephone conversation with a member of Enniscorthy's Chamber of Commerce revealed that there was enthusiastic support from some in the town for the exploitation of a greenfield site about a mile out the Dublin road. For what?

'Oh, probably a Tesco. The town needs it badly.'

Chapter 11 ॐ

| SENSE AND SENSIBILITY

The moral dilemma of supermarket shopping

'If these eggs are labelled organic, does it mean that the hens are cooped-up battery hens and just fed organic grain? Can organic grain still be genetically modified but it's just not on the label? Do the hens get to run around a field? Do I care if they get to run around a field? Do they care that I care?'

It's that time of the week that fills you with a back-to-school queasiness in the pit of your stomach. You've avoided it for several days now with trips to Spar and stocking up in On the Run when you get petrol. But you know in your heart ... time is running out. You've visitors coming and they might like to eat something other than liquefying lettuce leaves and potato salad with a carpet-like blue growth. The dog has a lean and hungry look and is taking an unhealthy interest in the baby's used nappies. For dinner last night you briefly contemplated trying to sell the kids on the virtues of sugar sandwiches. The fact of the matter is, unless you finally get up off your arse, grit your teeth and spend an hour or so of life's precious journey in the aisles of your local supermarket, a visit from social services may be on the cards.

Supermarket shopping has always had its irritations, but for many years, in the golden era before food became so complicated, the challenges were much simpler: it got no more complicated than searching the trolley bay for one that had four wheels that just might by chance travel in the same direction.

Those were the good times, the salad days of supermarket shopping. This was the golden period when none of us had heard of BSE, low GI diets or organic smoothies. Now we are bombarded with information on what foods are good, bad, life shortening, heart strengthening, brain boosting, joint lubricating and just about any claim an advertising company can come up with to make us pass money over the counter for their product. We are told to buy stuff that grows good bugs inside us to kill bad bugs. We are told that this vitamin found in this foodstuff will give you endless energy that will have you skipping your way into work like a schoolbag-swinging Shirley Temple. We are warned in scary voiceovers that *we are what we eat.* If we don't take the time to read labels in the supermarket and concentrate on what we buy, we are all going to hell in a handcart. Our insides will turn to the kind of sludge that may even end up frightening the toilet.

What should be one of life's simpler tasks has become a fraught experience. Like the Dashwood sisters in Jane Austen's creation, we are cursed with both sense and sensibility as we negotiate the supermarket aisle. On the one hand, we know we must employ Elinor's good sense, her practicality and restraint, even if it is somewhat lacking in romance. Elinor would get the job done, food miles, fad diets and latest scientific findings be damned. Our Elinor would be at the till rummaging for her loyalty card in her empire line dress while Marianne is still stuck somewhere around the organic fruit and veg.

Marianne, Miss Sensibility, is armed with too much knowledge about the way she *ought* to shop and has ended up reduced to a quivering mass of shopper paralysis. Supermarkets offer far too much choice for the Mariannes of this world and, likewise, for anyone who cares about food, worries about their arteries or isn't on mood-stabilising pharmaceuticals. Buy local, buy low fat, buy low salt, buy organic, buy fair trade, buy cheaply, buy sustainably! What is poor Marianne to do? She swoons and falls into a crumpled heap of skirts beneath the 'Buy Two for One!' Cillit Bang promotion. And there she lies.

There is no Colonel Brandon to gather us up in his arms; we are on our own in our quest for love and happiness in the aisles. Pushed and pulled by a confederacy of retailers, marketing people and food scientists, the weekly shop is an ethical minefield. And one thing is for sure: it will take more than a vial of smelling salts and a few tender slaps on the back of the hand to bring us round.

GRANNY KNEW BEST

How did we ever manage before the food scientists came along? How did generations of Irish mothers cook healthy meals without the benefit of knowing the difference between polyunsaturated and monounsaturated fats? What a terrifying deficiency of basic food knowledge our grannies must have laboured under. Imagine not knowing the first thing about antioxidants, flavonols, plant sterols and probiotics, not to mention the chill shudder that grips your spine when you consider her inability to distinguish between the anti-mutagenic and anti-carcinogenic properties of phytochemicals when putting together a Sunday roast. Not knowing which way the prevailing scientific wind is blowing on trans fats should alone be grounds to drum any parent out of the PTA, if they still have the nerve to call themselves parents.

The last thirty or forty years of nutrition science have hugely and unnecessarily complicated our food shopping. Who on earth finds it simpler to shop for a balanced mix of beta carotene, citric acid and antioxidising fruit and veg rather than sticking a carrot, orange and head of broccoli in the trolley? And the food marketers are catching up (and sometimes overtaking) the scientists in these helpful new categorisations. How often have you picked a container up off the shelf to examine a label repeatedly screaming about the product's probiotic properties? 'Great, but probiotic what? Yoghurt? Milk? Goat's cheese?' No, *Lactobacillus paracasei*, whatever that is when it's at home in the lining of your gut. Bet your gran feels humbled now that she never gave a moment's thought to anybody's intestinal health when drawing up a menu.

Nutritional science has been twisted into one of the greatest illusions of progress of the modern age. We are rational and logical beings, therefore we place our faith in what science has to tell us. But that science is being mediated by an army of ad men and marketing execs to tell us what they would like us to hear, people who are very skilled at employing nutritional science to promote the nutrients in their product which confer longer life, or promote weight loss, or help you to retain the complexion of your sixteen-year-old self (ugh, no thank you).

Gullible eejits that we are, we haven't first asked the question, how did we manage to feed ourselves so successfully for the last ten thousand years? Agriculture has served us pretty well since we first started organised cultivation in the floodplains of modern-day Iraq. Hundreds of generations of Irish mammies got by without ever losing too many of their offspring to cardiovascular diseases linked to obesity or colo-rectal cancer. But blind faith in progress allowed us to shake off mammy's wisdom. The industrialisation of food production in the post-war period was progress. Substituting terminology like 'cholesterol' for 'red meat' or 'butter' was progress.

Except it wasn't really. There was no good scientific reason for blinding us with the lexicon of the dietician, but there was a very good political one. The rise of nutritional jargon is linked to a very seldom remarked upon row in the

United States. In the 1970s, the Senate Select Committee on Nutrition and Human Needs was wrestling manfully with the question of diet and certain chronic illnesses. They made a lot of sensible observations, like how when meat and dairy products were scarce during the Second World War, instances of heart disease decreased, only to go back up when there was no more rationing, and the trend has been ever upwards since then. With logical, if not campaign donation-threatening innocence, the senators urged fellow Americans to cut down on their consumption of red meat, milk and butter if they wanted to prevent coronaries.

But lo and behold, the cattle ranchers freaked out. The ensuing campaign of vilification from the wealthy ranching constituents of the committee members pretty quickly convinced them of the error of their ways. A compromise was sorted that allowed the senators to stick to their advice, but now, instead of 'cutting down on red meat', Americans were advised to 'reduce saturated fat intake'. The die was cast from this point on. Governments were not going to be allowed to tell their citizens to eat less of any particular food. Out went the dictionary of foodstuffs and in came a puzzling new lexicon of nutrients and food groups for fear of ever again offending a wealthy lobby group. Cholesterol was pronounced bad, protein was good.[1] Which is as unhelpful as it is wrong-headed, because it meant that a diet of hamburger patties and rib-eye steaks containing both cholesterol and protein was simultaneously killing and curing you. Why couldn't somebody just tell us how many steaks a week was good for us?

It's a bit counterintuitive, but the consequence of introducing science into the food debate has been to inject equally large doses of anxiety and confusion. How ironic that the more we worry about how healthy what we eat is, the more our health can suffer. Michael Pollan, the author of several excellent books on food and nutrition, calls us 'Nutritionism's Children'. We are the generation for whom food has become a source of worry rather than enjoyment. Thinking about food will no longer set us salivating because we are too stressed out about the consequences of nutritional science's latest survey results. 'Is four hundred calories of potatoes the same as four hundred calories from a lamb cutlet? Do I go low fat and no carb, or should I carry a calculator to count the glycaemic index of what I eat?' For instance, common sense about how much fatty food is good for you has gone out the window. Hands up (and be honest), how many people think that a diet absolutely free of fat would be better for you than one with even small amounts of fat? In the United States, one-third of people believed this utterly ridiculous proposition to be the truth.[2] Unfortunately there are so many low-fat or no-fat products out there that you could actually subscribe to such a diet and remove a nutrient essential to the absorption of vitamins from your food.

Freed from fear of shortage, we have created a whole new set of food-related anxieties about what we put into our mouths. The logical conclusion of where this is all going is a new food-related psychological disease, orthorexia nervosa—an unhealthy obsession with eating healthily![3] How modern is that? It hasn't yet achieved clinical recognition that would put it up alongside anorexia nervosa, but some psychologists say that they are seeing more and more patients doing themselves physical and mental harm fixating on food science's mixed messages.

WHAT MORE COULD A BODY ASK FOR?

Hundreds of subtle influences, like little pushes and pulls on packaging and repeat advertising, mean we are guided by sensibility over common sense when we shop. Milk is as good an example as any. Elinor Dashwood would probably always have drunk hour-old milk direct from the urn. Forced into a supermarket, she would doubtless opt for full-fat milk from the nearest dairy and move swiftly on. The choice is not so simple for us Mariannes. The junior Miss Dashwood loves the idea of getting milk from mischievous nanny goats rather than exploited and tethered old factory cows. Unfortunately, goat's milk is rather too much of an acquired taste for her. So why not try and improve her health by choosing fat-free instead of full-fat milk?

Now pay attention, Miss Dashwood, here's the science bit, but only for the next few lines. Milk is pasteurised by heating it to sixty-three degrees Celsius for thirty seconds. We do this because it zaps the bacteria that could cause brucellosis and it has the added convenience for the retailer of prolonging shelf life. Then we stick it in a centrifuge to separate the cream. What's left is skimmed milk. If you want semi-skimmed or whole milk you blend the cream back in to add about two to four per cent fat. Ta-dah! But pasteurisation, a process born out of concerns over food safety, has been hijacked so that milk can now be marketed as not one but three different products, each more expensive than the last. Now as you arrive at the milk shelves you have a range of whole milk, low-fat milk and no-fat milk supplied by multiple producers. It should hardly be a choice. You know what your better instincts are telling you to do. The dairy may as well have labelled the three cartons 'Milk for Lardarses', 'Milk for People Who Try (But Not Hard Enough)' and 'Milk for People Who Run Marathons, Eat Muesli and Adopt Kids from Cambodia'. In a culture that constantly reiterates the nagging 'fat is bad' message, the implication of choosing no-fat or semi-skimmed milk over whole milk is obvious. Except some scientists are now starting to wonder whether we should have ever tampered with milk in the first place.

In the mid-1970s, links between dairy products and prostate and ovarian cancers were first explored. The nutrition police went to work and the usual suspects were rounded up. Saturated fats went down for the crime, and when all their other misdemeanours were taken into consideration it didn't look like they'd ever see daylight outside of the exercise yard again. But more recent research has put skimmed milk in the frame. An eleven-year study among twenty thousand men found that all increased risks of prostate cancer attributable to dairy products were down to consumption of skimmed milk.[4] A study of eighty thousand women found that those drinking one or more low-fat milk products a day increased their risk of ovarian cancer by almost a third over those who had three a month.[5] Could it possibly be the case that we didn't know everything that was going on in the processes we were subjecting our milk to?

There are no hard answers, but certainly pasteurisation kills off a lot more than nasty bugs. Good bugs and enzymes which help absorb the nutrients in milk get bombarded too. One of the more sophisticated theories in circulation is that the stripping of fat from milk removed vitamins A and D. Without these vitamins, it is harder for the body to absorb the milk's calcium and protein, which, if left rattling around inside you, can become toxic.[6] Who knows? It is only a theory. The long and the short of it is that more research is needed. The good news is that it is currently taking place in the largest laboratory in the world. The not so good news is that we are the test subjects. One day, perhaps ten years from now, having gathered sufficient data, food science will be able to make its final pronouncement.

Low-fat milk is a great value-added product for the food industry; value-added in this context meaning that they've done more to it so we have to pay more for it. The benefit for the retailer is roughly twenty-five cents more per litre for no-fat milk.[7] The benefit at our end is a little bit less clear-cut.

Unfortunately, it is too late to turn back the clock: very few people would settle for the taste (and warmth!) of milk straight from the udder. It would also prove impossible to unravel a production and distribution system based on existing herd sizes and best-before dates. But nobody can say, hand on heart and without any reservations, that no-fat milk is better for you than its full-fat big brother. The guys who thought of marketing skimmed milk to piggyback on the tide of no-fat fanaticism could probably see no harm in it. The CIA call these laws of unintended consequences 'blowback'. That's what happens when you arm the Taliban to fight the Russians and twenty years later your guys end up getting killed by the same weapons. Food science has stocked our supermarket shelves with plenty of products that should, in theory, be good for us, only that theory takes decades to become scientifically proven fact.

THE POSH SPICE DIET

Once upon a time we filled our bellies in the (hopefully) enjoyable company of our families. Now food has to make us feel good about having bought it. We expect it to unravel the consequences of our unhealthy lives. We want what we eat to say something about who we think we are. We demand a lot of our food, and our relationship with it has only become more confused. Have you ever thought about driving twelve miles to a farmers' market for a free-range goose for Christmas—because it's nice to support local producers and cook a traditional meal—but you'd die before you'd eat anything with that much fat on it?

It used to be that food was either plain good or bad. Now there's a third option—guilt-laden. Supermarkets are full of choices, difficult ones. For the caring, sharing, thinking person, walking around an Irish supermarket is a bit like tackling a particularly complicated physical and mental challenge, especially as our media continue to confuse and bamboozle us about what we should or shouldn't be eating. Very little of the information we get about what we eat is actually professionally mediated advice from a doctor or dietician aimed at a person's particular circumstances or lifestyle. Not only is new information on food relentless, it is frequently patchy and contradictory. While one month we are told to eat low fat, low GI, gluten free, low on dairy, decaffeinated, organic, the next month there are newspaper reports claiming that fat, dairy, caffeine and all the rest are in fact quite good for you and, hey, help yourself. And while the business of buying normal food has got pretty complicated, it's child's play compared to the craziness of the weight loss and health food industry whose tendrils have crept into all sorts of buying decisions to which we never gave any thought before now. Diet fads change faster than the weather, and if our celebrity magazines report one day that Posh Spice eats only boiled halibut or mulched blueberries, is it any surprise that people's attitudes to food are so messed up?

Food has risen above its station. It's gone from something we put in our gob when our stomach feels empty to punching way above its weight, creating dilemmas in shoppers, fears in parents, hours of TV programmes and debate, and reams of newspaper columns. A trip to the supermarket brings all these elements together.

'Why does the label on this bread say low fat when it has more calories per hundred grams than any of the others? Why are these packaged carrots cheaper than Irish ones when they come from Holland? Should I be buying Irish ones? If I get these organic potatoes I will feel good about myself, but then will I feel bad again because they are shipped in from Israel? Does this mean that wanting to eat organic products is increasing my carbon footprint?' No wonder the aisles are full of people standing still looking confusedly at labels on the back of plastic packaging.

Towards the end of 2007, food scientists from Coca-Cola, Nestlé, Tate & Lyle, Cadbury, Kellogg's and Mars gathered in the Burlington Hotel in Ballsbridge. Between them, these companies had generated vast mountains of cash from peddling sugar and fat, but their reason for coming to Ireland was altogether different. The conversation in conference rooms and around the bar was all about food's new growth markets. Products plugged as 'better for you' would, it was anticipated, grow by twenty per cent over the next four years, and everybody was getting in on the act. Coca-Cola, who had made itself a global market leader without ever once professing concern for public health, was now conducting clinical trials in drinks that would promote healthier hearts and bones. Danone had started selling a yoghurt drink containing antioxidants that claimed to nourish your skin from within. There was a lot of talk about chocolate products that would aid brain function, and even some excited rumours about a chocolate that would either cure or stave off Alzheimer's.

If the locally produced, organically grown vegetable stand in your weekly farmers' market stands at one end of the food spectrum, the Food Technology and Innovation Forum in the Burlington was about as far in the other direction as you could go. One of the more eye-popping presentations came from Canadian manufacturer Natraceutical. Their product Viscofibre makes you think you've eaten more than you have. It will form a soft gel in the stomach, which will expand to give you the physical sensation of fullness. Then, they claim, it will slow the uptake of sugar into the bloodstream before triggering the production of hormones in your gut that tell your brain you've had enough. In essence it's a food that's not a food but can convince your body that it is a food.

The lecture titles and workshop names told you all you needed to know about the direction the companies that had sold us the products that made us fat in the first place were now moving in: 'How to Make the Health Trends Work for Your Brand'; 'Harnessing Scientific Studies to Retarget Health-Conscious Consumers'.[8] The gathering was essentially about tapping into and further fuelling the consumer belief that we can eat ourselves to a healthier state of being. And it's a huge market in Japan, where the fortified or functional foods trend started (where else?), a market that is now worth well over €10 billion annually.[9]

While all foods are functional, the term 'functional food' is generally only used for a food that is considered to provide benefits over and above the nutrients required for normal health. Thus we have the supercharged omega 3 egg, the heart-saving dairy spread, the healthy gut-promoting yoghurt drink and so on. Initially, functional foods were aimed at Japan's ageing population, who were hitting bedtime just when health-care costs were going through the roof. Among the first to appear were useful drinks containing high levels of fibre. Such was their popularity that they were followed up with more high-tech foods containing health enhancers like oligosaccharides, sugar alcohols, peptides, glycosides, isoprenoids, cholines and others that the Japanese probably didn't realise they needed until they were told they did. After a quarter of a century of innovation in functional foods, the Japanese can now buy yoghurts that claim to make their skin paler and chocolate that is supposed to reduce stress levels.

We are several steps behind the Japanese in this evolution, but we are getting there. Some of the functional foods that appear on Irish supermarket shelves has proven beneficial effects on health. But some producers have gotten a little bit ahead of themselves, like Kellogg's All-Bran Flakes, which claimed to be 'Detox in a Box', a claim they have since removed.[10] Some offerings, such as low-fat or fat-free dairy and milk products, have consequences that we are only now learning about. The problem is that when we see these foods advertised and hear the health claims that are made for them, we feel that they should be part of our weekly shop, even though what we already have in our trolley is probably more than adequate to meet our needs. And while many of these foods have been found to have quite concrete beneficial effects, the difficulty is separating people who actually need them from general consumers who are fooled into thinking that they should be buying them for their family's general health.

One of the most impressive areas of growth has been in the yoghurt drinks that claim to improve intestinal health. Just look at the amount of shelf space given over to them. No doubt they are of some benefit to some of the people who drink them, but it is hard not to be cynical about some of the claims made for them. A study by the consumer magazine *Which?* suggested that the evidence in favour of probiotics was 'patchy'.[11] The Director of Ireland's Functional Foods Research Centre, Professor Gerald Fitzgerald, acknowledged that the sector had been damaged by a 'laxity of rigour and clinical verification of the claims made'.[12] In 2008, the manufacturers of Actimel, the French giant Danone, were rapped over the knuckles by the advertising watchdog in Britain. Actimel's TV ads said that its claims were 'scientifically proven' and invited you to go and look at their website if you didn't believe them. Unfortunately for Danone, their website said that access to the information on the scientific studies was limited to health care professionals. This has since been rectified,[13] but it might not help their case in the US, where the same product has been joined in a class action suit demanding the return of hundreds of millions of dollars to US consumers. The claim, completely rejected by Danone, is that they have used 'unsubstantiated' science to justify thirty per cent price hikes over regular yoghurt products.

In a 2008 Irish study of twenty-four food and drink items packaged as 'healthy choices' for children, all of them were found to be high in sugar, salt or saturated fat. Labels such as '100 per cent natural ingredients' and 'great for growing kids' were how some of these items sold themselves, but under closer (very close) examination, they turned out to be foodstuffs that in fact were pretty much the opposite. Treats for kids are something we expect to contain a lot of salt or saturated fats, but the items studied were far from bags of crisps: they were packaged portions of cheese, hams or spreads specifically marketed as healthy options for kids' lunchboxes. Mothers are consumers who are particularly vulnerable to a product telling them it has health benefits for children. In the survey, ninety-four per cent of mothers who responded said that health was the main factor driving what they chose for their kids' lunches.[14] By 2010, new European legislation should tighten up the health claims made by some products on their packaging, but in the meantime manufacturers will continue to court consumers who believe that when something says it is good for you, it actually is.

So as we take our kryptonite-enhanced yoghurt drink from the shelf and put it into our trolley, has anyone asked whether, in general, the technologising of food has been in the consumer's interests? It's not too much of a stretch to suggest that with functional foods the industry is only putting back in a lot of the goodness they have spent the last fifty years taking out, and charging a lot more for it. The Irish Nutrition and Dietetics Institute takes the view that if you eat a balanced diet including lots of fruit and veg, there's not much need for any of these products.[15] Food has become technologised, tracked, researched and commodified to within an inch of its life. So has how we buy it, how it is marketed, how supermarket aisles are arranged to trick us and treat us, and also how we then go home, throw some of it in the bin but still come back for more. Have we become far too willing victims to a whole world of stuff that we don't need to eat?

WHAT'S A GAL TO DO?

If we were all a little bit more Elinor and a little bit less Marianne, one of the first things that we could do is to shop around the edge of the supermarket. Everything you need to live, and live very nicely too, can be found in the aisles around the perimeter of your supermarket. Veg, fruit, dairy, meat and fish are traditionally in the first and last aisles you encounter. There is really no need to traipse up and down all the other aisles. Apart from salt and pepper, spices and olive oil, you could do quite happily without any (oh, okay then, with very few) of the cereal bars, cereal boxes, bags of polished white rice, confectionery and processed carbohydrates among which you meander for the majority of your shopping trip. Much of what is on these shelves forms what some call the 'Standard American Diet: Cereals Refined and Processed'. Now there's an interesting acronym.

If modern commerce has displaced much of our traditional diet, and our new diet is making us fat, unhealthy and undernourished, the remedy is pretty obvious. The less processed, the less 'value added', the less yet-to-be-proven food solutions we place in our trolleys, the better off we will be. If all that the last three decades of nutritional research and advice have done is make us feel miserable and anxious about our food, then maybe it is the nutritional advice that we need to cut out of our diets. There are signs that things are already moving in that direction. Some elements of the functional foods market in the UK have stalled: consumers confused by too many products clamouring for their attention may simply walk away from that section of the store. But the ever-fickle hand of media advice might also have pulled them in the direction of yet another healthy choice, the so-called superfoods. Perhaps the mania for blueberries, pomegranates and the like will also wane in time, but who can doubt that eating a real piece of fruit is better than eating a plastic potted fruit-flavoured yoghurt?

Ultimately a little bit more of the kind of common sense that won through for Jane Austen's girls in the end would do us no harm. Would Elinor listen to somebody who claimed to know that a single food product would lengthen her life? Poppycock! Would she seriously believe the advertisement that tells us, 'buy this and you will live longer'? Longer than what? How do you know how long I am going to live? How do you know that my gut is unhealthy or my liver fatigued? Dieticians can point us towards the foodstuffs that are guaranteed to shorten our lives, but claiming to know what will lengthen them is pure mumbo jumbo. A hundred years ago, did we suffer from cancers, heart diseases and diabetes in the numbers we do now?

Among the many things that changed in that time was that we became very sophisticated at improving upon nature to make her 'healthier'—while also improving profitability for some. Winding back the clock on much of that development, while cherry picking the best that the modern retail system has to offer us, would leave us occupying a very happy space. The good news is that it requires no raging against the machine, no people's revolution. We can put the wheels of our own quiet little revolution into motion every day. You could peel an orange at two-fifths of the cost of a litre of Squeez for your RDA of vitamin C. Some fruit juices helpfully add fibre nowadays, but for about fifty cents a week you could simply load up on a bowl of porridge. Why be persuaded to pour a bowl of breakfast cereal with added folic acid when you can get all you'll need from a few spears of broccoli, at about one-tenth of the cost and with none of the cereal's added salt? Milk with added omega 3 costs about twice as much as ordinary milk, but why wouldn't it occur to us to eat a tin of sardines once in a while? Benecol or Flora pro.activ may do the advertised job of lowering your cholesterol, but so does a raw carrot a couple of times a week. The ingredients are all there. We just have to resist our inner child, attracted to the bright shiny new toy, and pick from the foods that mammy knew were best for us.

Chapter 12 ∾

SQUARE PEGS INTO ROUND HOLES

Rejoice! Marks & Sparks have pre-peeled oranges

Have you ever tried to eat a chocolate bar while sending a text? At some point you have to put one or the other item down. The bar will take two hands to unwrap, and in the case of a Kit-Kat can't be broken easily with just one hand. M&M's can be poured directly from the bag into your mouth, but there's still all that fiddly opening in the first place. And crisps? Well, you can just forget crisps altogether; they've always been a two-handed operation, and they'll leave grease all over your phone. While you may think that this sounds like the stuff of schoolyard conversation, these are actually the debates at the cutting edge of where food is going. Because the food company that comes up with a range of snacks that don't require packaging and can be eaten while shuffling through the library on your iPod will have conquered food's newest frontier: the One-Handed Portable Snack as Meal Replacement.

According to one academic study, by the year 2030 most of us will spend as little as five to fifteen minutes a day in meal preparation.[1] Survey after survey has found that when it comes to food, we appreciate convenience over all other considerations, even flavour. And the food companies are there beside us, facilitating and even encouraging these choices. Cheese in strings because putting it in a child's sandwich takes too much time. O'Brien's sandwich fillings in a plastic container, presumably for Atkins devotees who don't want the bread. No-fuss, no-muss meals for people who think they have no time. Our retreat from the kitchen has been an enormous boon to the business of food processing and obviously a trend that is worth promoting rather than discouraging. Which came first? Did we start skipping breakfasts, then Kellogg's developed a breakfast cereal in a bar, or did we skip breakfast at home because we had a stash of bars in the car?

This is the area of the vast global food market that the industry loves. The more processing that goes into the product, the more value that is added and the greater the eventual mark-up. Why on earth would you want to sell somebody a banana for forty cents when you can sell them a 'Wholegrain Wheat and Other Cereal Bar with Real Dehydrated Banana Chips' for several multiples of forty cents? But therein lies the conundrum of making food more convenient. It can be annoyingly resistant to the idea of being ground, frozen, dehydrated, preserved and made non-perishable. Meats and fruits are pretty stubborn about how long they want to go on before changing colour. Wheat can be strangely reluctant to take on the flavour of honey or chocolate without a fair bit of jiggery-pokery. As a rule of thumb, it would be fair to say that the more convenient the food, the more time will have been spent modifying it in a factory. It's not unlike trying to shove a square peg into a round hole. If you push hard enough the peg can be made to fit, but at what cost to both it and the hole?

Spare a moment's sympathetic thought for the oranges unfortunate enough to be snapped up by the buyer from Marks & Spencer. Some marketing genius came up with the wheeze of peeling oranges and repackaging them in transparent cellophane wrapping. This is not a joke. This is something that M&S actually sold in Ireland. *Reductio ad absurdum.* Some bright spark product development manager thought that peeling the orange added value and offered consumers a better product than the trees in Seville were already capable of producing. After all, isn't orange peel kind of ugly? Hey, guys, we can improve this orange thing! No probs! We can just about imagine the meeting where this all happened:

Fearghna surveys the conference room. Today he is wearing his new suit from Hugo Boss, which he is sure will have the new girl from marketing licking his shoes by the end of his presentation.

'OK guys,' he addresses the room. 'This new product rocks, yeah?'

He clicks to the next page on his PowerPoint screen.

'Our target demographic is the Skippy (School Kid with Purchasing Power) and Woopy (Well-off Older Person).'

He glances at the assembled company to make sure they are on message with his favourite new acronyms.

'These segments of the market are ideal for this product because they have either . . .'—he looks at his notes— '. . . either dis-improving or, em, yet to be acquired manual dexterity skills.'

Uisneagh, who hates Fearghna ever since his golf handicap surpassed his own, throws his pen on the table and links his hands behind his head.

'Yeah, okay, there's, like, potential there, Fearghna, but isn't it a bit niche?'

Fearghna manages to keep a scowl from his features.

'No, not for what I'm thinking. Like, we also have the core demographic, the Lombards (Lots of Money but a Real Dipstick). They, like, love the whole deal that some, like, peasant in Belarus was paid, like, eff all an hour to peel an orange for them. It's kinda appealing, yeah?'

Uisneagh smiles and frowns at the same time.

'Yeah, like it. Now you're talking. What's our mark-up?'

'Sky's the limit on this one, Uisneagh. However, if it's, like, sensitively priced, the novelty value of this product could probably make some inroads into the Tepids.'

'Tepids? Now you're losing me.'

'Tastes Extravagant, Pay Inadequate, like, em, Dipshits.'

The unfortunate reality of all this creative endeavour, combined with all the peeling and repackaging, was one of the saddest spectacles ever placed on a supermarket shelf. An orange in its skin is a beautiful think to look at. A peeled orange with greying pith hanging off withering flesh is not. Encasing it in a rigid cellophane wrapper gathered into a Vicky Pollard ponytail at the top and held in place with a ribbon didn't improve matters either. At a distance they looked like what a tribe of Borneo head hunters might try and flog to gullible tourists. Close up they looked … well, sad.

'Bad example,' you're thinking, 'because Marks & Spencer always produced ridiculously over-packaged food.' Fair enough, they do take the art of unnecessary packaging to dizzying new heights. The real point here, though, is that food itself isn't wildly suited to Henry Ford-type mass production and marketing, yet we have tried relentlessly to push it into that round hole. The Marks & Spencer 'pre-peeled orange' is symptomatic of how our modern system of food production places a higher value on marketability, cheapness, uniformity, consistency of supply and ease of processing than it does on the product's intrinsic values. Once it sprouted from a tree somewhere in southern Spain, the orange ceased to be an orange and became an economic unit. From that moment on, the goal of the food production machine was to bring it to market as cheaply as possible. It became like any other commodity, from steel to microprocessors. It would be shipped to where it was most in demand and would fetch the highest price. Clean it, peel it, add stuff to it and you're into a whole new profit margin. And if next year it can be produced for less somewhere other than Spain? Well then, so be it.

WINNERS WRITE HISTORY, LOSERS READ IT

The conventional story of food production, the accepted version of events, is that modern food is humanity's crowning glory. Rather like a Soviet-era film showing happy workers shovelling grain into hundreds of lorries on a collective farm, we are told that our modern-day food environment is the result of a slick and sophisticated synthesis of science and mass production that has revolutionised our lives.

More food is being extracted per acre or per animal than ever before. It is safer, cheaper and more abundant than at any time in ten thousand years of human agriculture. We have a dizzying array of the finest, the most nutritious and delicious foodstuffs. With proper regulation and control we can supposedly check not just the background of every cut of meat, but the progress of each ingredient of a foodstuff through every stage of production, processing and distribution. Now food science is poised at the boundaries of a brave new frontier promising everything from a healthy and happy old age to an end to famine and food poverty, alongside the tantalising prospect of having whatever crop we want at whatever time of year our little heart desires.

Think of all the twentieth century's achievements: unravelling DNA, trans-global flight, mass communication, mastery of the atom. We are encouraged to think that these pale into second place alongside the liberation of one half of humanity, admittedly the wealthy half, from the daily drudgery of dishing up three square meals a day. How much more quickly society has advanced now that we are free from the tedium of baking bread and churning butter. How many more hours can be devoted to higher pursuits since we succeeded in relegating 'food' to a once weekly sweep up and down the aisles of Tesco? Women can take their rightful place in the workforce, men can swap the fields for finance houses and children can grow into healthy productive adults freed from debilitating malnutrition-related disorders. In two generations since the end of the Second World War, when Europe was on the verge of famine, we have progressed further and at a faster rate than at any time in human history.

Thanks to intensive agriculture practices and the efficiencies introduced by food conglomerates, we were freed. We may still be frail humans with a range of unpleasant bodily functions, but hey, at least we no longer have to spend the majority of our waking hours up to our necks in dirt in order to feed our puny bodies. This was the most complete social and economic revolution of the twentieth century. We were transformed from food gatherers to individuals free to pursue whatever economic, cultural or social project took our fancy. A silent, bloodless upheaval of the existing order, it required no referendum, there was no proclamation at the door of the GPO, the troops remained in the barracks. This dramatic change in the way we lived our lives was welcomed by popular acclamation. Had Karl Marx not been so fixated on the workers' chains and thought a bit more about their stomachs, he might have been on to something a bit more enduring.

It would be churlish not to admit that a lot of this version of food history is true and, yes, we are so much the better for it. But like all 'official histories', there were a few facts that had to be excised, like a couple of dodgy-looking cousins airbrushed from the family album. Take, for instance, the animal so finely engineered to produce more of itself that it can no longer stand under its own weight; the vegetable with all the appearance of its ancestors but none of the taste; the systematic exploitation of migrant workers, farming families living in abject poverty; or the horrendous slide into mental and physical incapacity of the variant CJD victim. Just as long-distance air travel has contributed to heating up the planet and the internet stands accused of dissolving our social glue, modern food production's triumphs are far from absolute.

'Nonsense!' you cry. 'This is the ranting of open-toed sandal-wearing malcontents.' Perhaps it is, but ask yourself—if the triumph of the food scientists and processing multinationals is so complete, why do we have such an unhappy relationship with food? Why do we find a trip to the supermarket so confusing? Why does a visit to the fridge leave you feeling guilty? Why, if the produce gathered under the roof of your local supermarket represents the pinnacle in human achievement, is so much of it just not good for us?

FEAR AND LOATHING IN THE FOOD AISLES

Most supermarkets are laid out with a level of deviousness that would leave Niccolò Machiavelli in open-mouthed admiration. While we think we are just moving from aisle to aisle simply picking up a few items, there are hundreds of subtle signposts telling us what to do and influencing what we buy: end of aisle displays that suggest a product is on sale or special offer when in fact it is just nearing its expiration date. Placing higher-priced products at eye level. Arranging crisps, dips and fizzy drinks all in the same place to encourage impulse buys. 'Shelf shuffling', the constant rearrangement of products to keep you there longer, increasing the possibility you'll buy something new. The supermarket chains have put years of research and millions into honing and perfecting their craft. While most of us cotton on pretty quickly to why milk is always the product placed furthest away from the exit, it's far from easy to see the tricks that are played with the food itself.

In a cleverly laid out supermarket, the smell of freshly baked bread will be the first thing to waft under your nose as you enter. Making a choice is as much a tactile experience as one of smell as you roll the still oven-warm loaf around in your hand. In most cases that very loaf will have been mixed and the bread part baked off-site before being rushed to the market and into an oven to create the illusion of a bakery in the community. Though the vast bulk of what we buy makes no such sophisticated claims for itself, bread making became industrialised in response to our demand for a cheaper, whiter, fresher product. Proper bread making takes time, however, and today's high-speed product only looks like a loaf of bread thanks to the addition of several new ingredients.

Fermentation of flour and yeast is what gives bread much of its flavour, but this is a process that inconveniently eats up lots of time. No problem; chuck in some salt, lots of it. Then add some fats to stop the bread collapsing. No, not just any old fats. They need to be fats that won't melt out of the bread as it is being baked, so the industry chose hydrogenated fats. Unfortunately these create trans fats in the body, which lead to heart disease. Palm oil was added as an alternative, but that too has been linked to increased risks of heart disease. Air and water are two of bread's principal (and cheapest) ingredients, but the air is only held in place by emulsifiers, which these days are made from petrochemicals. Yum. Then as a final touch to ensure longer shelf life before moulds start to grow, some brands will be sprayed with anti-fungal agents. For the supermarkets, bread has proved annoyingly resistant to the time constraints our economic model of food production places on it. That just won't do, so instead, bread is hammered into a round hole that better suits both supplier and consumer. The irony of all of this ingenuity is that anybody can make bread with just three simple ingredients. Sorry, four simple ingredients: flour, water, yeast and time.

Just after the bread aisle you will often be encouraged to get the boring old fruit and vegetable shopping out of the way before you move on to spending money on the real value-added goods. Have you ever noticed, though, that before you ever arrive at the first head of Cos lettuce you will pass a chilled cabinet with a range of washed and bagged salads? Why buy the ingredients for a salad when you can buy a ready-made one? Obvious, eh?

Until about fifteen years ago, bagged salad was a pretty unfamiliar notion, but currently in the UK it's a product that sells in the same quantities as breakfast cereals.[2] Putting lettuce leaves in a bag is as good an example as you will find of how we make food do something that it doesn't want to do. Left to its own devices in the open air, cut lettuce will begin turning into a slimy black mess within four or five days. In the excellent *Not on the Label*, food writer Felicity Lawrence first excited debate when she highlighted how dipping lettuce in chlorine and sealing it in modified atmosphere packaging increased its shelf life by fifty per cent. If clever manufacturers suck enough oxygen out of the bags and replace it with carbon dioxide, they can extend the shelf life by up to a month. The unfortunate but unsurprising downside is that the lettuce begins to lose its nutritional value. Vitamins C and E are depleted and there would appear to be fewer antioxidants in lettuce that has been bagged for as little as three days.[3] This loss of nutrients also occurs in leaves left out in the open. The difference is that your eyes tell you that the leaves in the modified atmosphere packaging are as good for you as the fresh head of iceberg a little further up the aisle, but they aren't. You have been conned.

What a master stroke bagged salads are! Take one iceberg lettuce and add enormous value by splitting it up into a dozen or so bags and selling each bag at the same price as a single head of lettuce. In the early days of the bagged salad miracle there were some teething problems; all that handling by badly paid workers who hadn't washed their hands after their last toilet break could lead to E. coli contamination.[4] That was resolved by giving the lettuce a chlorine bath before bagging it. Chlorine is an amazing chemical that will continue to kill bugs long after it is first applied, and that is what the surface residues left on some of your bagged salads will do. While some chlorine compounds are known to cause cancer, there is no proof that those on bagged salads will harm you. Nor, unfortunately, is there a large body of science saying that they won't.[5]

While your mixed leaf rocket salad in a bag makes no pretence of looking like nature intended, that is precisely what much of the rest of our fruit and veg aisle is trying to convince us of. Take apples, for example. While they look like pictures of physical perfection, mass-produced apples have become an exercise in anti-climax. Glossy, firm to the touch, just the right shade of green to trigger a slobbery Pavlovian response, unfortunately they rarely fail to disappoint. The first bite into the insipid soft tissue never delivers on the message your eyes have promised you. Boring, boring, boring! They are also substantially less nutritious than they used to be. You will have to eat three times as many apples as you would have needed forty years ago to get the same amount of iron.

Retailers know that we humans have evolved into shopping with our eyes. That's why they demand that producers go to extraordinary lengths to create a homogeneous community of Stepford apples. An apparatus called an 'intelligent quality sorter' will take up to seventy pictures of each apple as it passes through a 'singulator'. These can break down the preferred amount of red and green surface area by percentage. The manufacturers boast that their device has been made for a market in which 'buyers increasingly prefer products which have been sorted by external characteristics'.[6] The supermarket sets down exactly what shade of green, the degree of shine and the size of apple it wants to the nearest millimetre, and the sorter discards everything else at the rate of twelve pieces of fruit per lane per second. Blemished, bruised, scarred or worm-eaten apples won't even get their chance to impress you, as they will have already been discarded or at best pulped for juice. Vast quantities of perfectly good apples are wasted every year because they don't fit the very narrow definition of what the supermarkets know we want to buy.

In order to achieve the desired size, apple growers must cut back their trees and heap fertilizer on them. Then they must spray again and again with calcium to prevent blemish marks.

One of the few producers on this island lucky enough to get his apples onto the shelves of one of the multiples here says that over-fertilization reduces flavour. Speaking on condition of anonymity, he says that left to his own devices he knows he could produce a more nutritious and much tastier apple, a point he proves with a blind taste test contrasting fruit taken from his commercial orchard with a single tree closer to his house. There is no comparison. Smaller, blemished and looking more like a lemon than an apple, the 'home-grower' just explodes in the mouth. The flesh crunches as you crush, each chomp bursting over the sides of the tongue—it just feels right. The apples he sends to the supermarkets are grand, but that's all you could really say about them. It seems ridiculous to have to give a fruit and veg producer the cover of anonymity, but like many, he will only talk off the record because he believes supermarkets will 'de-list you for nothing'.

With as much as forty per cent of his produce not passing these beauty tests, he is contemplating availing of EU grants to cut down his trees. Imagine that the Common Agricultural Policy has come to this, paying Irish farmers to step aside for imported produce because we have a weakness for buying with our eyes rather than our nose, taste buds or brains. We seem happy to ignore the let-down of the tasteless, watery tissue under the immaculately presented skin.

Once again, the food industry has applied its considerable ingenuity to finding ways to make a fruit do things it doesn't want to. They will be uniform in shape, size and colour. They will meet our exacting demands for a product apparently free of imperfections, and they will last longer on the shelves before passing a sell-by date. The bizarre compromise we accept in return is that they are not as tasty and not as good for us as they could be, not to mention that while there used to thousands of different varieties of apples, now there are barely a half dozen on any store's shelves.

It's not just apples that are subjected to the food scientists' beauty parade. While the search for ever more disease-resistant, travel-tolerant, longer-lasting tomatoes that are uniform in shape and size has had success, not as much attention has been paid to what is in them. A tomato bought while Brian Cowen is Taoiseach will contain twenty-five per cent less iron than one grown while Dev was in office. The 2009 tomato will have thirty per cent less vitamin A and seventeen per cent less vitamin C than one grown during the Lemass years.

Tomatoes you could expect us to get wrong; after all, they have to be bred tough-skinned to get here in the first place. Potatoes, on the other hand, we have no excuse for getting wrong, yet today's potato is an entirely different beast from those grown just a generation ago. One hundred per cent of the vitamin A, fifty-seven per cent of the vitamin C and iron and over a quarter of the calcium have disappeared from our spuds in fifty years of ever more selective breeding.

After all the taxing and stretching of your culinary imagination that is required by the fruit and veg aisles, supermarkets like to spoil you with a little reward. Right about now you will probably come across what some in the trade refer to as 'in-home meal solutions for single-person eating occasions'. Only a complete Norman-No-Mates would buy for a single-person eating occasion, so not surprisingly they're sold to us as the more socially palatable ready meals or TV dinners. This is where the supermarkets can make some real cash. And such treats they have for us, too. The Tesco Value fisherman's pie contains as much as nine per cent fish. Their lasagne contains a whopping twelve per cent beef, which could be either Irish or British; they don't say which. The spaghetti Bolognese, which you would have thought might be equal parts spaghetti and Bolognese sauce, is actually twenty-six per cent pasta and thirteen per cent Irish or British beef and tomato purée. Hmm! So what is in the other three-fifths of my meal solution? Onion, mushrooms and carrots? Yes, a bit. But if we were making spag bol for ourselves, how many of us would reach into the cupboard for the pork gelatine? Or for cornflour? Corn by-products are a staple ingredient of ready meals because they are starchy and a great way of bulking up and thickening out meals, very often replacing more expensive ingredients like, oh, spaghetti or Bolognese sauce.

Just as with bread, a particular kind of fat is needed for the manufacture of most ready meals. Animal fats would work, but they are not remotely as cheap to produce as the oils produced from crops like soya or rapeseed. Cheap, yes, but also irritatingly unco-operative, these oils were reluctant to oblige by behaving the way they were required to. Over eighty years ago, some clever people found out that if you heat these oils to two hundred degrees Celsius and keep them on the boil for several hours while blasting them with hydrogen gas, you make them much easier to manipulate. For example, hydrogenation could be used to turn liquid vegetable oils into spreadable margarines. For a while we thought that this was all much healthier than consuming saturated animal fats, but unfortunately the manufacturing process had shot several decades ahead of medical research. About twenty years ago came the first rumblings that hydrogenation turned unsaturated fats into trans fats and that these fats were like an arrow made from lard shot directly at the heart. They should have been good for us because they were unsaturated, but they actually contained lots of LDL, the bad cholesterol.

Slowly, very slowly, the supermarkets started removing trans fats from their products, but there is still a lot of them about. Tesco, for instance, has removed them from their own brand products, but you will still find high levels of trans fats in anything from 'light' chocolates to granulated gravy, but you wouldn't know it because there is no requirement to label trans fats. The gradual move away from trans fats is one of those heartening people-power stories. We vote with our feet and the supermarkets and food processors pay attention. But as they run away from trans fats, they are doing so in the direction of full-on artery-clogging saturated fats. The Food Safety Authority of Ireland in a recent study noted that the sale of products containing trans fats was on the wane, but they were disappointed to note that manufacturers were embracing saturated fats as an alternative.[7]

Low-fat yoghurt is another of those damnably awkward foods that just won't do what it's told. Low-fat products have obvious attractions, but once you take the fat out the yoghurt tends to fall apart and has to be glued back together again. That can be achieved relatively easily with a bit of gelatine or pectin. Fruit, of course, also has an irritating habit of going off, so it is much easier not to add any in the first place. Modified starches are a pretty convincing texture substitute for most fruits, and so what if they are calorie-rich but nutrient-low: can you imagine how expensive yoghurt would be if it could only sit on a shelf for two or three days before being thrown out? Since it's devoid of any actual fruit, there will of course be some additives to create the impression of fruit-like colours, tastes and smells. Unfortunately, the product you end up with, although low in fat, will frequently have more calories in it than a bar of chocolate.

And here's a helpful piece of advice on the labelling front. If you want a yoghurt with actual apple in it, buy an 'apple yoghurt', whereas an 'apple-flavoured yoghurt' might not contain apples, but some of it is made from material extracted from real apples. And—pay close attention now, here comes the clever marketing bit—an 'apple flavour yoghurt' is an entirely manufactured taste sensation that wouldn't know a real apple if it fell out of the tree and banged it on the head.

There is an obvious value to using additives and preservatives that cannot be overlooked. They protect us from getting food poisoning. It would be very hard to guarantee any kind of consistency and safety in mass-produced food whose ingredients are sourced from multiple suppliers if producers weren't able to turn to the E numbers. But the E numbers that do the preserving constitute only about ten per cent of the additives that are licensed for use. The other ninety per cent are there to fool the eye and taste bud. And we consume a rather alarming amount of this deception. Add up all the additives in biscuits, ready meals, sweets, drinks and so on and you're talking about anywhere between six and seven kilos of E numbers a year.

It sounds like a lot of trouble to go to, turning unpalatable starches and fats into facsimiles of something they were never intended to be, but it's all about the bottom line. We don't know how much the supermarkets make in profits each year, largely because one of the largest players, Tesco, won't say, but turnover in the sector is worth over €15 billion. And it is the most highly processed foods (or, in supermarket speak, the highest value-added foods) that make the most money. If only the equation was as simple as we give them money, they feed us and make a profit in return.

Two-fifths of Irish adults are overweight. One-fifth are obese. Three hundred thousand Irish kids are overweight and obese. That number is going up by ten thousand kids every year. The two thousand premature deaths a year that are attributed to our diet, and the treatment of a range of other diet-related illnesses, costs the taxpayer about €4 billion a year.[8] The World Health Organization believes that sixty per cent of deaths around the world are related to changes in the way we eat and 'increased consumption of fatty, salty and sugary foods'. These guys are not exactly a bunch of alternative-lifestyle hippies prone to exaggerating the benefits of lentil soup, so when they suggest that a third of heart disease is caused by 'unbalanced nutrition' and a third of cancers could be prevented with healthier eating, you know that they are understating the case if anything.

To be fair to the food industry, it is we who decide what we put in our snack-holes. We are free to choose whatever we like and most shops will have the ingredients for dozens of healthy, balanced meals on their shelves. We just have to put them in our trolleys. Perhaps the reason we don't is because we are not as free to choose as we think. Rewind the clock for illustrative purposes to the early 1960s. Imagine yourself trying to make an informed choice about whether to start smoking. Bombarded by positive images of what cigarettes would do for your health and your image, and in the absence of any clear message about consequences on the packaging, why on earth wouldn't you? Fifty years later, look at your trolley and ask yourself why it is (in all probability) filled with the most heavily advertised products, and not less recognisable alternatives? Who among us reads the labelling, and when we do, can we make head or tail of it? Freedom of choice in food purchases is a bit of an illusion. Diet is probably at the root of as much illness as tobacco, but the food industry disputes as unproven the link between their products and our illnesses.

BRINGING THE MARKET TO HEEL

In 2007 we were all happy little Celtic Tigers. Property developers were our pin-ups and we thought it was only a good thing that the financial sector oxygenated the life blood of the economy with easy credit. In 2008 it all went Pete Tong and overnight we became born-again socialists, outraged that the government hadn't regulated the market before irrational exuberance crumbled into credit crunch. After decades of being a proper little bunch of Thatcherites, resistant to any kind of nanny state control, when somebody stole our ball the first thing we did was to seek protection in the warm embrace of her apron. What a turn-up for the books that some of the biggest boys in the schoolyard now wanted Nanny to come and supervise playtime.

Ronald Reagan's and Margaret Thatcher's credo that the markets would work best if freed up from state regulation was mirrored in food production. Producers were given a free rein and gradually the state retreated from offering nutritional advice. This vacuum was filled with advertising, with nobody to arbitrate the competing claims. This laissez-faire state of affairs has found its perfect expression in our government departments and quangos. The Department of Agriculture distributes the cheques and implements Brussels policy. The Department of Health is there to look after us when it is already too late, and only allocates the tiniest proportion of its budget to its Health Promotion Unit. The Food Safety Authority's primary concern is bugs in the food chain. The National Consumer Agency doesn't give two hoots what we buy, only whether we can get it cheaper in Newry. So who is there lobbying for change in how foods are made? In short, nobody with government backing. Who will chide the supermarkets for promoting their 'value-added' goods over cheaper, healthier alternatives? Who will nag us into doing the right thing and offer us helpful advice on how to do it? The system doesn't allow Nanny a role, but perhaps we need her.

As an agent of social change, the rise of convenience food rarely gets a mention alongside the pill or television, but it has been just as big. While we have been happily rutting away and slobbing out on the sofa for the last thirty years, our food habits have engineered a quieter social revolution. We have been transformed from households where cooking involved a full working day every week for one person to now, where it is barely half an hour a day. The benefits of being freed up have made our lives much more interesting, and it would be in nobody's interests to suggest a wholesale return to 1950s diets. But surely it's not too rose-tinted to admire a time when children were educated by their parents in the butcher's on what cuts represented the best value and how to ask for advice on their preparation? Are the communities and relationships that are created when you source food locally not preferable to bumping trolleys with strangers in the frozen food aisle? What family wouldn't benefit from preparing and eating a couple of meals together each week? Far from being a problem that requires a 'meal solution', food can be a familial glue, an agent of good health, and just plain fun. There's no need to cut supermarkets out of our lives, but challenging their dominance will force them to change in a way that would be better for all of us.

Chapter 13 ∾

MYTHBUSTING: A TALE OF TWO MARKETS

Are we too posh for Lidl?

Funny how a recession changes attitudes. It used to be that getting caught coming in or out of a discount superstore like Lidl or Aldi was the kiss of social death. Their pitched roof warehouses with aisles the width of a motorway lane were where 'others' went to do their shopping.

Headlines like 'Are we too posh for Lidl?'[1] seemed to accurately reflect popular opinion. There was an undercurrent of xenophobia in the way the quality broadsheets referred to the German interlopers and an implicit suggestion that you mustn't really care about the quality of what you put into your mouth if you were prepared to shop there. Well-heeled people didn't darken their sliding doors unless it was to avail of one of their non-food-related amazing special offers: fly fishing kits, scuba diving gear and a comprehensive range of power tools including every drill bit known to engineering. One Saturday afternoon in 2007, the car park of Lidl in Finglas was strewn with the packaging from a special offer on horse-riding gear. It seemed the horsey set had no problem buying bridles, headcollars and riding boots from the German discounter, but they wouldn't be caught dead with the packaging in their 4x4. Things have changed, though. Where once they broke out in a rash just driving past one of the German discounters, proud foodies will now boast about picking up top-quality Parma ham at rock-bottom prices there. Range Rover Vogues with this year's reg pull up in the car park alongside the uninsured banger with the Latvian plates. The aisles are alive with the well-modulated tones of the middle classes, whether they are from Knocknacarra, Killiney or Roche's Point.

It's a pretty impressive achievement, because location has been found to be the primary reason people remain loyal to a supermarket. We value convenience above price and we tend to be innately suspicious of anything that appears too cheap. That said, something interesting happened that fragmented our shopping habits and persuaded increasing numbers of us to split our shopping up over a number of stores. The National Consumer Agency undertook a survey that showed Tesco was over fifty per cent more expensive than the discounters on a basket of forty-four goods.[2] Tesco responded furiously, claiming the NCA's methodology was flawed, but what was most interesting was how the Republic responded to the NCA's findings.

In the past, price comparison exercises had all proven to be a bit of a yawn for the Irish public. 'Great, so they're cheaper, but it's not going to make me change my habits!' This time something snapped, though. A survey carried out by Amarach Consulting a few weeks after the NCA published its study showed that almost a fifth of the population had indeed changed their shopping habits. A further seventeen per cent said they were considering switching their shopping to the discounters. What must have really got Tesco and Dunnes worried was that the greatest numbers contemplating doing the dirt on them and switching their loyalties were among their most affluent ABC1 customers.[3]

How is it done? How are these newcomers to Ireland doing it so much cheaper than everyone else? Surely somebody somewhere must be losing out in order for consumers to save this much money? If the farmers complain about already getting screwed by the established chains, these guys must be making them flip cartwheels. If, that is, they are buying Irish at all. Perhaps they are feeding us muck that will turn us into a nation of chubbies reaching for the fat pants? And if they're not shafting either us or the food producers, they must be screwing their staff to keep costs down? The answers to all these questions, while by no means black and white, are not what you might first think.

THE KISS PRINCIPLE

Our first mistake when we look at an Aldi or Lidl supermarket is to think of it as a supermarket. It is actually much closer to the grocer's stores that our grandparents shopped in. Our second mistake is to assume that products left on a pallet or in cardboard boxes in the aisles is a mark of a store being less upmarket, when it is actually a smarter way of operating and allows for enormous savings. Aldi (who are much more willing to talk about themselves than the ultra-secretive Lidl) probably wouldn't like to be thought of as a grocery shop of old. Their marketing stresses that they are the new generation of food shop, as radically different from a supermarket as supermarkets were in their day from grocers. But in a way their success is a step back to the future.

Like Aldi, your local grocer didn't offer your granny endless choice. If she wanted a pint of milk there was one kind on offer, not a thirty-foot display with several varieties of the same product being offered by numerous competing brands. She bought her milk happy in the knowledge that her grocer would do his best to assure the quality of the product. Aldi claim that they have in effect done the same thing. They say that on our behalf they have stood in front of the same display and researched which is the tastiest, most nutritious, best-produced and best-quality product. Then they sell it, and only it. As they say themselves, 'we do not believe customers need a choice of twenty types of tomato ketchup when we sell the best one in the market and have done the hard work for the consumer'. KISS: Keep It Simple, Stupid!

But looking at the prices, they must be selling us poisonously cheap crap? There are some processed foods in their range that, to put it mildly, look like a bit of an adventure. But one of the retail food sector's best-kept little secrets is that when you shop in Aldi, you are actually buying many of the same products sold by Superquinn, Tesco and Dunnes, but at a fraction of the cost. Aldi's meats come from Larry Goodman, the same plants that supply Superquinn. Aldi's own brand crisps are made in County Meath by the same company that makes Tayto and King crisps. Their tea and coffee are Bewley's and Robert Roberts. Their yoghurts come from a company in Clonakilty who also supply Dunnes, Tesco, Centra, SuperValu and Superquinn. And Aldi flour, which your intuition might suggest comes from a monstrous mill somewhere in the Ukraine, is actually Odlum's flour.

The list is considerably longer; in fact, forty per cent of Aldi's produce is sourced in Ireland. But suppliers fret that their primary brands could be damaged if it slipped out that their premium-priced product is exactly the same as the one on offer considerably cheaper in Aldi. So everybody stays schtum to spare their blushes. It's no act of altruism. Aldi believes that their sales have been boosted by stocking Irish products. When they arrived here their butter was sourced in Britain. They switched to an Irish brand produced by Town of Monaghan Co-op and sales went up.

It is as much a case of psychology as anything else, in the opinion of managing director Donald McKay: 'Irish product gives a reassurance to consumers on safety and quality.' Most of us would make an automatic assumption that quality is the first part of the equation to suffer when we buy food at prices as low as Aldi's. What has actually been removed is 'stores, people and logistics. We are stripping out cost, not quality,' says McKay.

From the first principle of keeping it simple flow all sorts of other efficiencies. If you only have nine hundred, as opposed to ten thousand, different products to source, pack, ship, store and stack, you need a lot fewer staff. A typical Aldi store will be run by roughly fifteen people. As Aldi expands to around one hundred stores, they will employ in the region of two thousand people. Dunnes Stores, with 156 shops on the go in 2009, employed 18,000 people. And that is where the savings are made. By removing the choice, or making choices for you (married to a few other simple bits of Teutonic efficiency), Aldi has changed the shopping landscape. What they have in effect done is to go back to the future. It is a supermarket with the element of choice removed. Ironic, isn't it, that it was Germany and not Britain, the so-called nation of shopkeepers, that came up with this backward-looking innovation? Britain's gift to Irish shoppers continues to be Tesco.

When Lidl and Aldi arrived in Ireland in 1999, speculation in trade union circles was that they would prove to be poor employers: if they are able to sell at those prices, they mustn't be paying their staff properly. Tales from the Continent of menstruating female employees being forced to wear headbands (so managers could allow them to take extra toilet breaks) did the rounds. Left-wing websites and blogs were alive with tales of exploitative rates of pay, but if such practices ever existed they haven't been adopted in Irish stores and the unions don't have much cause for complaint with either company. Aldi's rates of pay are very attractive for the sector. Working on the tills or stacking shelves, an Aldi store assistant can earn €2.20 above the minimum wage, and after four years with the company a store manager winning his or her bonus can expect to gross €65,000 per annum, leaving them roughly twenty per cent better off than their peers in other stores.[4]

Happy staff, happy suppliers and happy shoppers? If it all seems a bit too good to be true, that'll be because it might yet be. Having captured almost five per cent of the grocery market by the end of 2008, Aldi announced further expansion with an ultimate goal of around 160 stores nationwide. With this kind of purchasing power will come increased responsibility. Aldi will become king-makers in the Irish food industry. As they only sell one or two versions of any item, whichever supplier their gaze falls upon will have hit the big time. For instance, being Aldi's main supplier of soft drinks no doubt means a lot to the Gleeson Group, who also produce Tipperary bottled spring water. (Yes,

Aldi water is actually Tipperary Water: we were surprised too.) Think of what it would mean to the Gleeson Group if that contract went elsewhere. Local markets can also be skewed by a company like Aldi when they make deals for a chain of over eight thousand stores. The company that wins the deal to supply Aldi its cured hams becomes a pig-guzzling behemoth that dwarfs all other competitors in its locality. Aldi claim that unlike their rival supermarkets they are not looking to squeeze their suppliers for whatever bit of margin they can get out of them. They say that once they have made their choice of supplier, it is in their interests to see that that company thrives. If they are sincere in this, then they are truly introducing an entirely new set of values to the grocery retail sector. Who knows? Only time will tell, but as of the summer of 2009, IFA president Padraig Walshe pronounced the people who dealt with Aldi as 'happy because Aldi honour their agreements'.

In an ideal world there is a lot about Aldi that would be different. It would be nice if they intended to source more than fifty per cent of their produce in Ireland. It would be reassuring if Lidl were a little bit more transparent about the origins of much of their stock. Unfortunately, both companies will no doubt continue locating their shops on the fringes of towns where land is cheap and parking spaces plentiful. It is a business model which has unfairly sucked the commercial life out of town centres and should be resisted, not facilitated, by the planners. For a company that has the wit and local knowledge to realise that the crisp-eating sensitivities of the Irish are best catered for by selling them a home-grown champion of crisps, it's a pity Aldi's architecture is so alien to the Irish landscape. Low pitched roofs on warehouses that are vaguely evocative of an alpine ski chalet may look at home in Achselschwang, but they definitely don't in Arklow. And both Lidl and Aldi have some pretty rum products mixed in with otherwise good fare (everybody suggests staying away from the beans).

To dwell on these issues is to miss the bigger picture, though. They have achieved something remarkable in Irish consumers: behaviour change. Slowly we have been brought around to the idea that choice and plenty come at a cost and are probably unsustainable. Compare your basket after an Aldi shop and a Tesco shop. The difference is that as you leave Aldi, you'll have what you need, whereas exiting Tesco, you'll have a good few additional items that you want. Aldi won't sell us stuff that we don't need, simply because it doesn't help them keep their costs down. More and more of us are embracing the concept of having what we need and not always what we want.

MYTHBUSTING 2: PHILIP GETS HIS HANDS DIRTY AT THE FARMERS' MARKET

'Our customers are single parents, not judges' wives,' says Charles Ryan as he shovels carrots by the two-kilo handful out onto the Carlow town pavement. 'We want to sell to the housewife from the poorer estates. They're our target market. They can come to us and get a week's worth of fresh veg for €15 for a family of four.' Sprouts, cabbage, cauliflower, parsnips, onions, rooster potatoes. James and Charles Ryan from Kildavin have a 'pile 'em high and sell 'em cheap' philosophy at their stall in Carlow farmers' market. Since the downturn, their customer base hasn't increased any, but they find that each customer is spending more with them as they are now conscious of getting better value.

Robert used to spend €10 a week buying 'a few old bits and pieces' from the Ryan brothers. Since he was made redundant he now spends €20 a week at the veg stall, which goes a lot further than if he were spending it eating outside the home. In the queue after Robert, Catherine is annoyed with herself. She went into Superquinn in Carlow earlier in the day and began doing her meat and veg shopping even though she knew the Ryans were only over the road. She put a single parsnip on the weighing scales and printed off the price tag before recoiling in horror at what Superquinn were charging: ninety-three cents for a single parsnip. She put it back on the shelf. Charles Ryan hands her a bag of six. If he is tempted to chide her for even contemplating shopping elsewhere, he shows no sign of it. 'Two euros, please, Catherine.' 'Look, I even brought over the sticker to show you,' she says, showing it stuck to the back of her gloved hand. Charles forgives her with the smile of an indulgent parent as he loads up her bag with the rest of the week's groceries.

None of what she is saying is new to Charles, but perhaps it surprises him that it takes consumers so long to learn that you can't beat the Ryan brothers on value. Their carrots are two-thirds the price of Superquinn's equivalent. They'll send you off with over three kilos of potatoes under your arm for €2, where the supermarket's best deal on roosters is €4.19 for two and half kilos. As for the parsnips, if you can live with a bit of mud, the Ryans will sell them at a quarter of the price Superquinn is looking for.

So what guardian angels of the recession are helping the people of Carlow munch through the crunch? Discount importers? Rejects from the supermarket quality control line? Veg off the back of a lorry? None of the above. The Ryans are one of about sixteen stallholders at the farmers' market in Carlow every Saturday between nine a.m. and two p.m. 'Surely not a farmers' market? Overpriced organics and fashionable artisan foods? I'd spend less ordering food hampers from Fortnum and Mason.' It's true that farmers' markets have made a negative impression on many. For several years you couldn't open the pages of the *Irish Times* magazine on a Saturday without reading a glowing piece about yet another niche food producer selling direct to the public. Then the worm turned, and as so frequently happens, having inflated our opinion of farmers' markets to ridiculous levels, the media took it upon themselves to burst the balloon. The idea got out that farmers' markets were exorbitantly expensive, and that they were also impractical places to shop unless you can sustain yourself for a week on falafels, savoury crêpes and scented candles. They were an indulgent shopping experience, not a place to get your weekly essentials. Some commentators decided that the notion of farmers' markets just didn't sit with the orthodoxy of the new asceticism. Thus farmers' markets became a by-word for unsustainability and a prime example of rip-off Ireland. 'People obliged to count the pennies and drag their shopping home on the bus after a gruelling day on the minimum wage are not likely to take a jaunt to a farmers' market,' thundered the *Irish Independent* in 2007.[5]

One notably expensive and badly run market in Dublin had jaundiced a few journalists to the whole farmers' market concept. From there it took only a few more negative feature articles to write off the markets in terms of the value they could offer to cash-strapped consumers. 'I suppose a part of the problem is that people assume markets are exclusively organic, and therefore more expensive,' is Kildare sheep and cattle farmer Peter Byrne's assessment. 'There's a place for organics, and every market should have an organics stall, but it's not for everybody. The price of an organic fillet steak would make anybody feel poor. But when you compare like with like, we beat the supermarkets on price and quality every time.' It's no idle boast. Premium minced meat in the supermarket is €10.99 a kilo. Peter sells a tastier equivalent for €7.60. His stewing lamb is twenty per cent cheaper than that of his supermarket rival but has been 'hung for nine days and handpicked by me'. And he's not lying about flavour. One pound weight of Peter's lamb with some market veg and the bare minimum of seasoning made us the most delicious Irish stew eaten in our house for a long time.

Careful shoppers are wise to the value the farmers' markets have to offer. If you match like product with like, and not organic sirloin of beef from the market with a scraggy bit of diced meat for a stir-fry from the meat counter of a multinational, there is no competition. And nor should there be. The farmers' markets have few of the overheads of the stores. They should be able to offer you considerable savings on cost with every bit of the quality you'd expect. Perhaps negative word of mouth was generated by those who filled their baskets and compared the bottom lines without noticing that they had slipped in an organic honeycomb, a jar of loganberry jam and a loaf of durum wheat bread. This is the kind of stuff we never dream of adding to our trolleys during a normal shop. No wonder the resulting spend could be pretty shocking.

While the best farmers' markets seem to be run on a sort of co-op basis so that costs are kept down and genuine local suppliers encouraged, some markets (and more typically the ones closer to Dublin) have been blighted by the nature of their produce and the greed of their owner/managers. Some are traders masquerading as farmers, and traders' markets masquerading as farmers' markets have damaged the reputation of the sector. They buy in goods, usually organic, frequently from overseas and very often from the same producers supplying the supermarkets, and sell on at a price fractionally below that on offer in the stores. Essentially they are supermarkets without the light, heating and piped music. The greed of some of the operators of the privately owned markets has in some cases forced prices up. 'It would take you half the day' to make the money to pay the owner for the privilege of renting a stand at his or her market, says fruit grower Tom Malone. That said, there are a lot of good markets out there, and local authorities can do a great deal to encourage more.

The number of farmers' markets appears to have stalled at around 120, with a question mark as to whether that figure will hold steady. Typically stallholders can expect to take in about €450 a day, and it is reckoned that north and south the markets turn over about €28 million per annum. Pretty small stuff in the overall scheme of an all-Ireland grocery market worth about €17 billion. But it's important to those who have gone down this route. For Peter Byrne, farming three hundred ewes and forty head of cattle in County Kildare, the Carlow and Athy markets provide him with ninety per cent of his income. For the Ryan brothers, selling off the pavement beats dealing with the supermarkets any day. 'Six weeks' credit before the supermarkets pay up versus cash in hand here on the day. Which would you choose?' asks Charles.

When a market is properly run, like the one in Carlow, it is an interesting blueprint to an alternative future. Unlike some of the privately run enterprises that ask traders for up to €80 per stall per week, Carlow County Council in its wisdom recognised that this was a service that should be offered by local government, not private enterprise, to the citizens of the town. They gave the stallholders free water and electricity and left them to get on with it. There is only one operating rule: the stallholders must be producers themselves and sixty per cent of what they sell must be their own produce. This has kept out the traders masquerading as farmers. It also means that the vast majority of produce on offer is fresh, locally produced and seasonal. All highly laudable, but Carlow distinguishes itself on value for money. Selling eggs at €1.80 a half dozen (€2.15 in the supermarket), Tom Malone is one of the organisers of the Carlow market. 'They got it right here,' he says of the local authority. 'The right level of support without unnecessary interference.'

Farmers' markets won't change the grocery shopping landscape in Ireland, but if there were enough of them they would provide a healthy safety valve for food producers. Manners might be put on the more aggressive supermarkets and wholesalers if there was an alternative outlet for their harried producers. They might be less inclined to squeeze farmers so much on price if the farmer had a viable market in his or her area to supply instead. For the consumer there are the fringe benefits of knowing that at least twenty per cent of your money isn't being spent on unnecessary marketing and packaging. It's also a nicer way to shop. You simply can't beat the product knowledge of the farmer who has reared, butchered and is now selling you your steak.

Some moves are being made to increase support to farmers' markets, but once again there is a lack of cohesive government support. Officially the Competition Authority and the government see opening up the doors to giants like Wal-Mart as the route to true competition in the food sector, while they turn their back on the ready-made home-grown solution. Companies like Wal-Mart sell cheap because they achieve economies of scale by sourcing their food globally: there would be little or no room on their shelves for anything grown in Ireland. And if Wal-Mart were allowed to do it, Tesco and Dunnes would have little option but to follow them down the same route. Ignoring smaller indigenous industries in favour of multinational giants is a great short-term fix, but over the longer term it all too often proves itself to be breathtakingly stupid policy making of the kind that is now resisted by our European partners. Small and local alternatives are not a complete solution in themselves, but four years ago there was no Carlow farmers' market. Now hundreds of 'judges' wives and single mothers' have changed their shopping habits and are unbothered by picking a few muddy carrots off the pavement.

Replicating this success elsewhere is such a no brainer that … well, perhaps we'd be better served by letting Carlow County Council run the country?

Aldi and the farmers' markets occupy very different ends of the consumer spectrum. They may never dominate in any brave new retail world, but they do represent a very different dynamic between consumer and producer. The dominance of the supermarket model and all their smaller corner shop imitators has not served us well. Any new player who introduces a different, if strangely familiar, way of doing things has to be a welcome departure. It's not an ideal picture, but it's a move in the right direction.

Chapter 14 ∾

WHERE DOES IT ALL COME FROM?

There used to be lots of farming going on in the world, now there isn't

Currently there are 6.7 billion of us running around the planet, eating burgers, brown rice, olive tapenade, whatever takes our fancy. To feed us all there is about 1.5 billion hectares of arable land chugging away turning out food, and for decades we have lived (at least in the Western world) with food costing us less and less. This meant that we thought about it less and less, and expected the stuff to land in our supermarket trolley with the same availability and price that it always has. But by 2007, that was all about to change.

After decades of surplus and cheap food, a large spoke went in the wheel when world demand for food and energy began to outstrip supply. The era of cheap food was well and truly over. Suddenly your weekly shopping receipt was the price of a good suit. Fifty quid got you five items and a bottle of wine. What the hell had happened to cheap food? We blamed the supermarkets, and we blamed the consumer agencies for letting the supermarkets get too big. We blamed the farmers. It's all their fault; them and their brand new John Deere tractors. We blamed everyone but failed to understand that in the wider scheme of things, this was the end of a false era of cheap food; we had been hoodwinked into thinking things would always be this way, without environmental or social consequences. By abdicating ourselves from food production and not caring whose hands it went into, we were writing our own history as the people who forgot about food.

'The 21st century will be characterised by a desperate struggle for sufficiency in food, energy and water, rising food prices in real terms, extreme hunger and famine as well as social unrest and wars over resources.'[1] This unpleasant scenario was recently painted not by a doom merchant NGO, but by a farmer, Michael Murphy, who keeps dairy cows and has agricultural interests in the US and New Zealand, countries where large amounts of the Western world's food is produced. It's a view of the future which is unpalatable but increasingly accepted as being the case. After a period of world food surplus, we now have a world food shortage, and the shortage will increase unless we change some fundamentals in our systems for producing food.

A combination of factors led to the rising cost of food. In addition to speculation on commodity prices, one of the reasons for price hikes in food is something far beyond our control, which is in the long term likely to get significantly worse. In 2007 and 2008, extensive droughts hit the large food-producing areas of the world. The US, Australia, New Zealand and Russia all experienced large crop losses. Some areas were particularly badly hit: Russia's southern Rostov region declared a state of emergency as severe drought devastated 1.4 million acres of its grain-producing land. The fear of running short of food became a reality for thousands of people around the world. It was an idea especially shocking to those of us in the West, who had for generations taken food for granted and viewed shortages as something that would only happen in Ridley Scott films where aliens landed and sucked up all our resources.

Climate change and the effects of global warming were suddenly beginning to be felt. Seasonal droughts and the flooding of low-lying regions are becoming more intense and more frequent. Environments that previously tended to swing back to their normal state after climate aberrations weren't doing so. Between 1975 and 2006, China lost one-quarter of its land to encroaching desertification; not a good scenario if you've two billion mouths to feed. And it's not just a problem confined to remote, far-away places. The Irish Climate Analysis and Research Unit (ICARUS) at NUI Maynooth found that climate change and water shortages will hit some regions of Ireland heavily, particularly the south-east, where growing some crops such as potatoes will be unviable in the future. Instead we shall see fields of maize and irrigation pipes criss-crossing our fields, perhaps even a few sunflowers and a vineyard for good measure.

While it may appear that Ireland will become more gloriously Mediterranean in appearance, the larger picture is a pretty alarming one. Put simply, climate change is resulting in less food being produced throughout the world, particularly where formerly rich agricultural land is turning to desert. This means that food prices are unlikely to fall for any of us, and at worst, food shortages, famine and social unrest over the availability of food (already seen in the 2008 food riots of Egypt, Haiti and Bangladesh) will become more frequent events.

SMILE, IT'S THE MILKMAN

Another factor driving food prices ever upwards is the huge growth in the world's population, currently growing at an alarming 75 million per year. It's a population that is now living longer and is increasingly urbanised. For the first time in history, 2008 saw the proportion of the world living in urban areas reach fifty per cent. The real significance of urbanisation is the shift in eating patterns that accompanies it; people who move from rural to urban environments change from a diet heavy in starchy foods, largely carbohydrates, to one containing more meat and dairy products. We have seen this trend develop strongly in China and India, where Western-type eating habits are advancing at a heady pace. By 2010, China is expected to have around 250 million middle-class Chinese. Their changing dietary penchants for dairy produce and meat have brought about some of the price rises in these commodities. And while millions move away from peasant farming to industrialised jobs in both China and India, the numbers farming staple products such as rice and wheat will fall. Farming will have to change to a much more industrialised model if these countries are to feed themselves, but given new patterns of eating, this has big implications for the way food is produced across the world.

In China particularly, demand for beef and dairy goods is spiralling. On a visit to a dairy farm last year, the chirpy Chinese Premier Wen Jiabao said his 'dream' is for each Chinese child to consume a pint of milk a day. Products from the Chinese dairy giant Mengniu (the official milk of the Chinese space programme) offer to 'fortify the Chinese people', with packaging showing a space-suited boy clutching a glass of creamy goodness.[2] While it is boosting their own milk production, China remains a huge net importer of dairy products, its domestic appetite far surpassing what it can produce itself.

And lo and behold, extraordinary demand sometimes leads to extraordinary events, as evidenced by the scandal over baby milk products in China in 2008. Some dairy producers had been watering down milk. This kind of scam was regularly tested for, so they added a protein called melamine which would mask the effects of the added water by artificially elevating the milk's nitrogen levels. This adulterated product (for you would hesitate to call it milk) found its way into everything from baby formula to chocolate bars. Thousands of babies endured the excruciating agony of passing kidney stones. The World Health Organization, a body renowned for mincing words and pulling punches, said that it was 'not an isolated accident, but a large-scale intentional activity to deceive consumers for simple, basic, short-term profits'.[3] Cue much grovelling by the Chinese authorities and promises to beef up their regulations.

The scandal resulted in the withdrawal of products from twenty-two companies and has damaged the credibility of a country which is seen by many to have no cultural ties with dairying and milk production. It wasn't unfamiliarity of the product that led to these fatalities, but greed on the part of two brothers from the Hebei province who confessed to mixing the melamine into their milk.

The scandal created huge mistrust of Chinese dairy products, so much so that in the aftermath of the events, Starbucks, which has over a hundred coffee outlets in China, replaced its dairy creamers in its outlets with a soya alternative. When demand gets this big, as is the case in China, the rush to supply it frequently overshadows safe practice. China's economic miracle will most likely create further food scandals in years to come as more corners are cut and consumers suffer.

And in the background, the environmental and more hidden costs of this huge rush to produce food and produce it fast will have long-term impacts not only in emerging economies such as China and India, but across the world. If everyone now wants to drink milk, huge numbers of dairy animals which weren't there before suddenly need to appear and start producing, fast. In order to do this, you need megatons of grain to feed the cows. This is also the case with beef cattle: eight kilos of the stuff is needed to produce a kilo of finished beef. Most beef across the world is mass produced in corrals where the animals eat grain-based animal feed. Ireland's grass-fed cattle, roaming about with joyful abandon, are an increasingly unusual lot.

So, in most of parts of the world, that nice filet of beef on your plate or cool glass of milk has taken shedloads of grain, and thus masses of arable land, to produce. Between 1980 and 2007, cereal production across the world increased by an average of two per cent a year, while the increase of feed use has averaged over 3.5 per cent a year, so the grain you see growing in fields around the world, which could feed the poor, is often going down the gobs of hungry cattle that end up in burgers for middle-class kids in Hong Kong and

Dublin. Before the collapse of milk prices, Irish farmers thrived from the race
to supply China with food, and Irish enterprises such as the Irish Dairy Board
have seen sales of skimmed milk powders, chocolate and dairy drinks expand
rapidly in the new affluent China. Other companies such as Glanbia and
Dairygold are reaping the substantial financial benefits of what some are
calling the Klondike of the East, while canny competitors, such as Kerry
Group, were in at the early stages, acquiring food businesses in the Asia Pacific
as far back as 1994. Kerry Group now has three manufacturing plants within
China, with sales of $425 million in the region for 2007.

RESOURCE HUNGER

This change in eating patterns throughout Asia has major implications for the
price of grain and the future demand for it. Not only has meat itself a huge
appetite for grain, it's also pretty handy at using up water at an astonishing
rate. In drought-ridden Australia, one of the major beef-producing nations,
they have found that seventy thousand litres of water is needed to produce
one kilo of beef. Water is becoming a resource under threat, and in India,
which is suffering increasing shortages, rice farmers use around five thousand
litres of water to produce one kilo of rice.[4]

> By 2025, 1800 million people will be living in countries or regions with
> absolute water scarcity, and two thirds of the world population could be
> under water stress conditions. The situation will be exacerbated as rapidly
> growing urban areas place heavy pressure on neighbouring water
> resources.[5]

Some seventy per cent of the world's water is used in agriculture. As our
population grows, we need more water to grow more crops to feed that
population, particularly more meat and more cereals for feeding meat. And so
it continues; our appetites for particular foods and an exploding demand for
dairy and meat products further stress our very basic resources.

With the rising demand for burgers across the world, finding land to rear
cattle has been a big issue in South America, where large swathes of the
Amazon rainforest have been felled to produce beef animals. So instead of
sucking up nasty carbon with our trees, we are producing yet more
greenhouse gases in the form of methane from cattle and the particularly
pernicious nitrous oxide. According to the UN:

Livestock are one of the most significant contributors to today's most serious environmental problems. When emissions from land use and land use change are included, the livestock sector accounts for 9% of CO_2 deriving from human-related activities, but produces a much larger share of even more harmful greenhouse gases. It generates 65% of human-related nitrous oxide, which has 296 times the Global Warming Potential (GWP) of CO_2. Most of this comes from manure.[6]

Maybe that burger's not looking so good now.

China itself is treading its way through an environmental minefield. With further climate change approaching, melting glaciers and the prediction that the Yellow River will seasonally run dry, the Chinese Academy of Sciences said that 'the melting glaciers will ultimately trigger more droughts, expand desertification and increase sand storms'. Increasing soil erosion and the spread of deserts on arable land lead the institute to admit the situation is an 'ecological catastrophe'.[7] Around the world, droughts in 2007 and 2008 brought decreases in agricultural production, despite the big changes in climate and temperature that were forecast not as yet being upon us. In 2007, Ukraine, Russia, Argentina and China had to ban wheat exports in order to keep enough for their own populations, and many Asian countries have followed a similar path with rice.

The rising price of oil has also meant that land has been turned over to growing energy crops, such as corn maize. In the US, usage of corn for ethanol grew from one million bushels in 2000 (five per cent of the corn crop) to six million bushels in 2007. An OECD paper published in 2007 suggested that the notion of biofuels as a 'champion energy source' may be misleading when we examine their cost and environmental impact. 'The combined cereal shortfall in Europe, North America and Australia in 2006 was more than 60 million tonnes—relative to the 17 million tonne increase in cereal use for bio-ethanol production seen in those countries.'[8] Are we starving people in the developing world because of our desire to drive greener cars? It's a little more complicated than that, but however you look at it, grain that goes into biofuels is not going into the food chain, and it's a development that is here to stay, with Shell saying that the bio crops we are growing today are an important bridge to the environmental saviours, the second generation of biofuels, which are at this stage five to ten years away.

WHAT ON EARTH CAN WE DO?

Another big factor driving food prices upwards was the speculation on commodity values by hedge fund managers sitting in comfortable leather chairs in London. Although at present their leather chairs and the figures on their computer screens are likely to be giving them one giant pain in the ass, foodstuffs, known as 'soft commodities', were part of a bull market that had been on the up for the past ten years. As world population grew and the China/India giant kept getting hungrier, betting that grain would go up in price in six months was a fantastic earner for the hedge fund boys. The price of these commodities was a one-way street until things reached a tipping point in late 2008, when commodity prices fell, coinciding with the banking crisis across the world. The commodity balloon had burst, although some view its collapse as only a temporary phase. The USDA's chief economist, addressing Congress in May 2008, drew attention to the fact that the 'futures market prices suggest that grain and oilseed prices will remain high over the next few years', adding that 'tight supplies will keep markets volatile with much attention paid to growing conditions worldwide'.[9]

After all, populations across the world are still rising apace, climate is messing about with yields and unless we all cut down on what we eat, it's clear that demand will not lessen. What's also alarming those who follow the supply of food across the world and issues of food security is that the 1.5 million hectares around the world that produces what goes into your gob today *is not going to increase in size*. In 2008, global stock levels of grain fell to their lowest level in twenty-five years. In fact, it is estimated that by 2050, the amount of land under agriculture will not have changed from its present level, while the population will have risen to 9.2 billion. So what are the extra two billion people going to eat? Where is it all going to come from? We're not yet at the point of a Hollywood disaster movie plot, but food supply is set to become one of the defining issues of the next thirty years unless we:

1   beat all hedge fund managers to death with sliced pans,
2   enact a Malthusian-type plague to kill off half the world's population, or
3   get a handle on growing enough food for the world without allowing its status as an economic commodity to result in shortages and starvation.

Still, almost one billion people (of whom 845 million are undernourished) do not have enough to eat. Seasonal famines and longer food shortages are in the present day not just commonplace but frequent occurences. The sad thing about it all is that we have been here before. Why is it that we fail to learn the essential lesson that is pretty obvious about producing food—that if we hand over the control of the stuff completely to private enterprise, speculators and multinational corporations, we lose control of it? By relinquishing our control of food, have we relinquished our control on life? Yes, of course it's easier to let Tesco make your lunch for you than make it yourself, or buy an M&S sandwich in the afternoon that was flown in from the UK this morning. In our hearts we know that this is all a bit unsustainable, and that the price of that convenience is something that has made us vulnerable to not being able to afford it in the future, or, in the very worst case scenario, to not getting enough to eat.

The level of *remoteness* that food now has, the distance between us, the eaters, and where and how it was produced, has been a hallmark of the way we have consumed and traded food for the past few decades. The price of oil and transporting food, together with yield shortages which increase the costs of staple ingredients, will force consumers to look more closely at what we eat and why the hell it's costing so much. And there's nothing that affects human behaviour more than price. If carbon trails across the sky don't do it for us, the prohibitive cost of that packaged sandwich flown from Leeds might make us switch to one made down the road instead.

It doesn't help that our distance from the basic realities of food have led us to treat it with little regard and to throw much of it in the bin. We don't have figures for Ireland, but over in the UK a third of all food purchased is wasted, according to WRAP, the Waste and Resources Action Programme.

> Researchers found that more than half the good food thrown out, worth £6 billion a year, is bought and simply left unused or untouched and that £1 billion worth of wasted food is still 'in date', costing local authorities £1 billion a year to dispose of.[10]

Are we as bad as this in Ireland? It's truly shocking stuff. Apparently, stopping the current rate of food wastage in the UK would be the carbon-saving equivalent of taking one in five cars off British roads. Put that in your BLT sandwich and eat it.

WHO'S RUNNING THIS THING?

Organisations such as the World Trade Organization and the EU (with its Common Agricultural Policy) are meant to be in control of the whole show.

Their job is to govern trade agreements and subsidies in farming and food production across Europe and the rest of the world. Apparently.

The WTO, viewed by many as a vast, impenetrable planet of lobbyists and carpetbaggers, was engaged in a round of talks to reach an agreement on farming and trade since 2001. Finally, in 2008, they failed to reach a conclusion. The Doha Round, as it was known, was launched with great hoopla in the Qatar city of that name nearly ten years ago in an unrecognisable era when oil was $30 a barrel and food mountains were an acceptable, if guilty, reality. The Doha Round promised that this was the era to give the developing world a better deal, allowing more tariff-free penetration of their foodstuffs and goods into the well-upholstered, fat-padded West.

Representing the interests of Irish farmers, the EU negotiator for the WTO talks was the former Northern Ireland Secretary Peter Mandelson. Scandal-soaked Mandy, who has twice had to resign from Blair's cabinet, became the EU's commissioner to the WTO in 2004. His aim over the course of the round was to secure an agreement that was, unfortunately, deeply unattractive to European and particularly Irish farmers. The main problem was that he proposed allowing beef imports, largely from South America, into the EU which would hurt Ireland's beef industry. Irish farmers protested, and finally marched on Dublin in advance of the Lisbon vote, warning that if Ireland didn't get a veto on WTO, they would all vote a Big Fat No to Lisbon.

Panicked about possibly ending up locked out of the European party, Brian Cowen described the WTO deal as 'unacceptable', and guaranteed Irish farmers a veto on the most unpalatable aspects of the deal on beef. The farmers put away their placards and went home happy, but in the heel of the hunt, it looks like most of them voted No to Lisbon anyway. After all, the IFA's No to Lisbon campaign had scared everyone blue for months about how ruinous the agreement was going to be for Ireland. By the time the organisation changed its tack and muttered, 'Actually lads, em ... let's vote yes to Lisbon', it was too late. And while Ireland wrestled with the Lisbon fiasco, Planet WTO was beginning to panic.

As the summer months of 2008 limped wetly on, time was running out to reach agreement in talks that had lasted nearly a decade and were looking shaky in terms of actually signing off on anything. Not only did Mandelson have angry European farmers to deal with, but new opposition entered the ring in the form of Carla Bruni's beau, French President Nicolas Sarkozy. As France took over its leadership of the EU, Mandy and Sarkozy began a war of words over Europe's position on the WTO deal. Sarkozy attacked Mandy's plans to free up global markets and to cut EU farm subsidies, saying his farm proposals were a madness that would cost Europe a hundred thousand jobs and destroy the farming and particularly the beef sector. Mandy retorted with the caring, sharing response that cutting EU subsidies benefited poorer, developing countries. Mandy sulked, telling the press, 'I'm not going to be bullied',[11] Carla strummed her guitar and French farmers gathered in cafés, smoking.

Meanwhile, in South America they were spitting mad. Viewing the bitch fight in Europe and potential failure in signing an agreement, an Agence France-Presse article reported how 'South American leaders slammed "outrageous" new EU immigration laws as xenophobic, and charged that farm subsidies were a key cause of the global food crisis. Argentine President Cristina Kirchner warned, "Not long ago, nobody could imagine the food problem would mushroom so swiftly, with situations that take you back to the Middle Ages and people dying over a grain of food, or a crust of bread."'[12]

But Sarkozy wasn't listening. In his mind, the opportunity to save European farming, and do something for trade, was via the long-planned review of the EU's Common Agricultural Policy, which falls under the term of the French presidency. Every year Brussels spends €44 billion, or around forty per cent of its budget, on Europe's farmers. Many of them are French. Over in the USA they were about to publish their equivalent in the shape of the Farm Bill, a paper that would set the subsidy regime for the next five years, and one that was mired in disputes, with the White House occasionally threatening a veto if the bill resulted in raised taxes.

But bills and drowned agreements aside, the rise in grain and rice prices continued apace, the reality being that speculation in the commodities market is beyond anyone's control (bar that of the speculators themselves). So, unfortunately, are the effects of climate change and the price of oil. 'What can we do about rising food prices?' people said. Looking to negotiators at the WTO or to the Common Agricultural Policy to bring about change in even the short to medium term was unfortunately not the answer.

OH MANDY, YOU CAME AND YOU GAVE WITHOUT TAKING

Barry Manilow lyrics aside, Mandy came, but couldn't give us or Europe what we wanted. The World Trade Organization talks failed in the dying days of July 2008. India and China, and their bargaining positions with America, were what chiefly brought down the talks, though European and Irish farmers still blamed Mandy while at the same time saying that the collapse of the talks was the best thing since sliced bread. After all, for Irish farmers, Mandy's deal had sucked from the start. When the talks collapsed, it was Irish farmers who were the most delighted guests at the party. Instead of facing the horror of tariff-reduced South American beef, we had a return to the status quo. It had actually been a close shave—the talks failed literally at the eleventh hour. Members were debating the nineteenth out of twenty outstanding issues that required resolution when an ongoing row between America and India brought the 153-member mêlée to a deafening close. Irish farmers breathed a sigh of relief. Padraig Walshe, leader of the IFA, breached security at the talks to congratulate the French agriculture minister in person on his defence of European farming, saying, 'The determination with which President Sarkozy and the French government defended European agriculture was greatly admired and appreciated by Irish farmers.' He called on the Tánaiste to support the French in 'tearing up' the EU WTO offer as the other parties had walked away from the table.[13]

After all, rules about how we trade food should not be incorporated into agreements on the tech sector and car parts, which only promotes food insecurity throughout the world. Others, like most of the developing countries around the world who are still suffering from being left outside or heavily penalised from trading in our rich markets, would disagree. It looks like bilateral agreements between countries will be the only way forward—countries of the developing world could have benefited more from trading en bloc. The collapse of WTO left Irish farmers happy, but in a world of food shortages, starvation and abject disparity, happiness can come at a cost.

AND WHAT DO WE HAVE HERE?

In the excitement of Lisbon and shaking our fist at the WTO, did we stop to ask the question, how important is farming to Ireland? Aren't we a food-producing nation? Then why the hell is the stuff so expensive?

The extent to which Ireland is, or is not, deeply involved with producing food is a key issue that tends to fall into the unspoken gap between protesting farmers and the city workers who walk politely past them, ignoring the flat-capped rabble. Was our espresso-carrying generation already light years distant and uncomprehending of the farmers' concerns, or do they feel that farming has a cherished place in both our economy and our national identity? Are we still a farming and food-producing nation or just a bunch of average consumers, tech workers and fast-food eaters? And if we're not the Emerald Isle any more, what exactly are we?

The exact importance of farming to the Irish economy is something that's hotly disputed. Irish farmers like to claim that without farming we are nobodies, and they point out that no indigenous industry has grown to match farming's contribution to the economy. Farmers are custodians of our environment and our rural heritage. Ireland *is* farming. Fighting the opposite view from the red corner, we have the non-farming-based economists such as the ESRI and anyone the IFA claims is juggling the figures to make farmers look less important. In 2008, the ESRI's medium-term review said:

> Twenty-five years ago the agricultural sector was still a significant contributor to GDP and agriculture exports were 12% of GDP. However the last quarter of a century has seen a transformation in the structure of the economy. Today, agriculture only accounts for 2% of output.[14]

This kind of announcement puts farmers into a spin. It degrades their importance and dilutes their negotiation positions in the WTO and CAP reform. But over in the blue corner and warming up for a fight are the Teagasc economists. Immersed in the agricultural sector, these boys put agriculture's contribution to the economy at eleven per cent 'and growing'. Clearly there's a tussle going on over how important agriculture actually is and whether or not we still see ourselves as a farming nation.

Full-time farming in Ireland is in decline: on eighty per cent of farms, the farmer and/or spouse earns some source of off-farm income, whether from employment, pension or social assistance. There are still some big earners, particularly in the specialist dairy and tillage operations, but the fact remains that in 2007, over a third of Irish farms had a farming income of less than €6,500.[15]

It's these sorts of figures that have prompted many in Ireland to pack in farming for good. Most, however, still point out that if they choose to keep their hand in by farming on a part-time basis—a couple of sucklers out the back of the house, or a few ewes up on the mountain—it's the way of life that keeps them coming back, not the money. Hardly surprising. Most Irish farms have cattle as their dominant enterprise, with dairying occupying second place at twenty-three per cent. Farms in Ireland are on average still small: thirty-six per cent are less than twenty hectares, and about twenty per cent over fifty hectares. In terms of how farming is going across the world, these figures don't make for good reading. Big is beautiful. In fact, big is not the word, we're really talking about *enormous*. It's the large commercial farmers that are making headway in the sector. Most Irish farms are too small, too unspecialised and shrinking in terms of importance rather than growing.

Irish farmers claim that their poor incomes result from not being given fair prices for what they produce. Shortage of labour and the cost of labour, much of it now foreign, pressed many farmers into accepting margins on their produce that simply weren't tenable. Many stopped farming. Consumers, even Irish ones, didn't really care. It's a reflection of what has become, in recent decades, an unfortunate lack of empathy and a disconnection between the eaters and makers of food. In a world where the supermarket multiples control most of what goes into our fridge, we have got used to food being cheap. However, it's precisely this cheapening of food that has driven most farmers out of their traditional way of life and will ultimately result in us paying *more* for food in the future. The more power we put in the hands of the multiples, the more they set the bar on what food costs at farm level and at the supermarket till. Giving huge corporations the responsibility for feeding us mightn't prove the smartest idea we've come up with.

Farming is now a vastly complicated business and worldwide it is seldom about producing *good food* at fair prices and in the right quantities. In the late 1980s and 1990s, food surpluses and schemes such as REPS meant many Irish farmers had land that was set aside and not producing anything, a situation hard to believe in the present-day world food climate; but in farming, matters are never as simple as they seem. A shortage in one part of the world doesn't always mean that someone can jump in to fill that gap. The regime of subsidies determined by the EU means that farming has become so complicated it no longer resembles any kind of straightforward pattern of production. Many advocate a return to a simpler system, particularly those who see Ireland's food future in the production of high-quality, simple produce reared locally and eaten locally, such as grass-fed beef and dairy, and niche produce that yields healthy margins, such as organics and artisan foodstuffs.

Organics won't feed the world at large, they are not even what feeds most Irish families; our toasted cheese sambo or pasta salad is still produced, on the whole, by mass agriculture that is international whether we like it or not. And, yes, this is rarely good for the environment; even Irish farmers point out that you won't see hedgerows in the grain basins of the US, or butterflies or birds or locals working on farms. Much of the labour on America's super farms is from developing nations whose workers often suffer dreadful working conditions and union bans in order to earn a dollar. If this is the sort of system that produced cheap food in the past, and we're still prepared to accept it, so be it.[16]

However, for those of us interested in food and wondering which burger is better or best, there should at least still be a choice to buy local over mass-produced imported products. You just might have to look a little further than your local supermarket. It's a pleasant surprise that many farmers' markets are actually quite cheap, and that dividing your weekly shop between the supermarket and other, more local sources is not as difficult as it initially seems. Sourcing food from more varied sources than a multinational chain can be an interesting exercise which involves getting more involved with what we put into our bodies, where it comes from and what we're paying for it.

The future direction in which Irish consumers and Irish farming will go is deeply dependent on our attitude to food. If we are prepared to change our relationship with food, we will not only broaden the range of choice in what we eat, but also help our rural culture, our health and, in the long run, our wallet. Whether Irish consumers are prepared to do this is yet to be seen.

Chapter 15

CHICKENS WITH PAGE THREE BREASTS

At the limits of animal husbandry

I t could have been the plastic food wrappers. At the end of the day, it wasn't anything to do with them, but some thought they were the first sign that all was not as it should be. Department of Agriculture inspectors said that mixed in with the dough, bread, muffins and assorted pastries that had passed their sell-by date was the packaging that had enveloped some of them when they sat on the supermarket shelf. Had this been a simple case of out-of-date food being landfilled, this would have been nothing out of the ordinary, but the inspectors were looking at the raw materials for pig food. Bars, buns, loaves and sliced pans gone past their sell-by dates are all perfectly good for a pig. Indeed, it would be immoral to landfill food that could be recycled. But what if the plastic and paper they originally came in weren't being removed? It doesn't bear thinking about. A pack of sausages would effectively be Wrapper in a Pig in a Wrapper. A great headline perhaps, but it wasn't the case, and the truth was to prove a lot more complicated.

Wheat had become astronomically expensive during 2008 and recycled breadcrumb was a cheaper alternative in the manufacture of animal feed. Wasting not and wanting not, the Millstream Recycling plant outside Bunclody had made a business out of taking unsold bread from the supermarkets and unused dough from bakers, reconstituting it as breadcrumb and selling it to farmers as one ingredient of animal feed. Following the loose threads of an investigation triggered by elevated dioxin levels in a sample of pork, the department's inspectors visited Millstream on the first Tuesday in December. They told UCD food safety expert Professor Paddy Wall that they couldn't believe the amount of plastic they saw mixed in with the unsold bread awaiting processing.[1] The packaging was a digression, though. Disturbing as it was, there was no way that it alone could be responsible for the levels of dioxin that were turning up in pig meat. The company subsequently vehemently denied that they had knowingly mixed plastics into the feed and fuel oil turned out to be the culprit. Perhaps the inspectors were unfamiliar with the recycling operation, as the Department certainly weren't regular visitors to the plant. Perhaps if they had been better versed in what went on at Millstream, the plastic wrappers would have been recognised for the red herring that it was.

The Sunday morning following the unannounced visit to the recycling facility, RTÉ Radio mustered its troops for an unscheduled *Morning Ireland*. The cat was well and truly out of the bag. Irish pork was poisoned with dioxins at anywhere between eighty and two hundred times the permitted level. Rory and Monica O'Brien in Mitchelstown woke to a regular day on their pig farm. They should have been unaffected, as none of the contaminated feed had come anywhere near their farm. But by nightfall their world, and the entire Irish food industry, would be in turmoil as the government pressed the nuclear button and recalled all pork products. Not just every rasher and sausage, but every single product that contained as much as a gram of pig fat had to be taken off the shelves. As with the 390 other pig farms in Ireland that hadn't used the contaminated animal feed, the O'Brien's family business had been effectively closed down. What should have been a busy period in the run-up to Christmas had turned into a nightmare for Rory O'Brien. He told the *Irish Times* that while on one level the industry was contemplating 'meltdown',[2] farmers faced a more immediate quandary: if they weren't selling, where were they going to find the money to feed the thousands of animals in their sheds?

The dilemma for the government was one that might not have been considered even ten years ago. That they took the action they did is a measure of consumer anxiety about food security. All the available scientific advice was that even though dioxin levels were at two hundred times the permissible dose, there was in fact little or no risk to public health. You would ingest more dioxins if you ate a charred sausage off a barbecue. So why jeopardise the entire pork industry with a very high-profile product recall? Well, let's consider what would have happened if they hadn't acted with complete transparency. Somewhere along the line, something would have leaked to the press. How much worse a pickle would they have been in then? Such is the constantly simmering level of public anxiety about food adulteration or accidental contamination that nothing less than effectively shutting down the entire industry until it got its house in order was what was required to allay fears. It was never a public health issue; what was at stake was market sentiment, which had been severely bruised in recent years.

PIGS DO FLY

The blanket recall of pork took Irish supermarkets by surprise, but they swung into action pretty quickly. From Saturday afternoon, staff began emptying walls of fridges containing ham and bacon joints, rashers, sausages, puddings, pies and just about everything on the deli counter. Well-known products from Galtee, Denny, Ballyfree, Olhausen and Shaw's lay forlornly in green plastic crates, while the customers themselves milled about the empty shelves, confused, bewildered and worried. Staff were badgered with questions. Should I have eaten that BLT at lunch? What about my pepperoni pizza still in the freezer, is it safe? I'm pregnant, will my baby have dioxin poisoning from what I've eaten? In the first three days of the crisis, the Food Safety Authority of Ireland helpline received three thousand calls from consumers uncertain of what to do with their sausages.

It didn't help matters that the papers went mad with headlines more suitable to the *National Enquirer.* 'Poison pork panic: Irish pigs were fed on plastic bags' said the *Daily Mirror,* while the *Daily Express* ran with the headline, 'Shoppers told: Don't eat toxic Irish pork!' On newspaper websites across the globe, from *Le Monde* to Singapore's *Straits Times,* dioxins in Irish pork were the most commented-on stories. CNN even went so far as to draw parallels between the Irish pig crisis and BSE, bird flu and the contamination of baby milk in China.

By Monday, the EU reacted, requesting member states to 'detain pork meat and products from Ireland and to control for the presence of dioxin and dioxin-like PCBS' (polychlorinated biphenyls). On the same day, China banned imports of Irish pork and feed, while Singapore and South Korea suspended the import and sale of Irish pork products. Back at home, our pork processing lines came to a halt, workers were told to go home and customers queued up in supermarkets to return the tainted food. It was a disaster for Irish agriculture, but the reaction of other food-producing countries to our misfortune gives rise to the suspicion that politics and trade will always trump food safety.

The Chinese were still on the back foot as an upshot of their own melamine scandal, which had killed several infant children and hospitalised thousands[3] (see Chapter 14). But then the unfortunately named Mr Hogg in Bunclody handed the Chinese the high moral ground. Pretty much immediately the Chinese got busy assiduously testing all EU pork products for dioxins. A positive result would hand them the leverage to ban not just Irish but all EU pork products. A trade war is so much easier a solution than enhancing regulation and enforcement. 'As long as your produce is bad we don't have to worry so much about making our bad produce better.' Similarly, over in the Brazilian embassy, joy was unconfined—their cattle farmers had endured years of (entirely justified) criticism from Irish farmers about the quality of their beef.

The crisis exposed global food safety faultlines, but in the short term Ireland's name was mud. Crisis groups were formed to see if they could possibly weed out Irish pork products from the system; a big ask when every single part of a pig is processed except the oink. In Belgium, distributors of pig heads to the Orient sat gazing at shelves and shelves of similar-looking snouts from all over Europe wondering if it was possible to ethnically cleanse the Irish from their stock, or would the whole lot have to be binned? Salami makers in Italy may have produced their own pork, but were cursing the Irish because they sourced fat from us to bind their sausage together. The scandal broke with just two and a half weeks to Christmas, spelling disaster not just for the ham market but also for items as improbable as traditional suet for the Christmas pudding.

As consumers, we were all understandably confused. Wasn't there supposed to be complete traceability? You know, farm to fork and all of that? Why wasn't it possible to sort the contaminated from the clean? After two decades of food scares and political promises of reform, the dioxin scandal revealed pretty huge holes in the system. While a side of ham could be traced with confidence from farm to abattoir to butcher to your plate, the secondary processing of pigs had been deliberately obscured by the industry, perhaps because we would discover the range of unexpected products that pig ends up in. Or worse, if black puddings made in Ireland carried traceability labels, consumers would become aware that the pig's blood in this most traditional of Irish products came from places like Brazil; and that in some cases the pudding contains no pig's blood at all—it comes from cattle. It seemed that however sure we were that we knew where our food comes from, the pork contamination revealed big weaknesses in the system.

As it stood, Irish pigs were traceable from farm to factory, but once they were killed and entered the glamorous world of secondary food processing, our pigs racked up a lot of air miles. While beef animals have a pretty clear line of traceability, pig meat goes through huge amounts of processing: it's mixed with other pig meat sources and made into sausages, puddings and pies, and rendered down into crazy ingredients for foods that we hadn't even heard of. This means lots of different animal parts and animals are mixed together. Even before the crisis in Ireland, a Belgian food processor who provides fats to the manufacturing industry had noticed an increase in PCBs in composite samples containing pig fat from several member states. He was still trying to work out which country was causing the problem when the Irish recall was announced. Not a good way to have to figure these things out, is it?

Even the *Farmers' Journal*, always cautious about upsetting the apple cart during times of crisis, laid the blame at the successful lobbying of certain sections of the pork industry for exemption from full traceability.[4] While farmers had been subjected to intense scrutiny in every aspect of what they did, political influence had allowed processors at the other end of the industry to wriggle away from the same kind of inspection. Not for the first time, it took a crisis to highlight a problem and make the case for change.

SUZANNE HAS A FOOD SCARE DÉJÀ VU MOMENT

The uncomfortable truth that the food we eat might be contaminated with dioxins was unfortunately not an unfamiliar one to those working in the farming or food sector over the past ten years. In 1999 I spent a hair-raising month examining the background to Belgium's dioxin crisis, which we subsequently filmed for *Ear to the Ground*. What had started off as a small news story that Leonidas chocolates and other Belgian food products were to be withdrawn from sale turned into an incredible tale of greed, corruption and government cover-up of food chain contamination that eventually cost the Belgian economy $1.5 billion.

Things first began to go wrong in Belgium early in 1999, when chicken farmers noticed strange nervous disorders and premature deaths among their chicks, combined with a high ratio of eggs failing to hatch. Overdoses of antibiotics used in animal feed were initially thought to be responsible, but after a month or so, one producer of animal feed took the bull by the horns and sent a sample of feed to a lab specialising in the analysis of polychlorinated dibenzodioxins (PCDDs) and polychlorinated dibenzofurans (PCDFS).

The results weren't good. With a bit of detective work, a storage tank containing animal fat for processing into animal feed became the focus of attention. The tank belonged to Verkest, an animal fat-processing firm in Flanders. They had sold fat to eleven animal feed producers in Belgium, France and the Netherlands, with the resulting feed containing about seven hundred times the permitted concentration of dioxin. Eventually two executives from Verkest were arrested and tried for fraud and falsification of documents. During the process of investigation, no public warning was given and thus no national or EU recall of Belgian food products took place. Consumers were completely in the dark.

It was not until June of that year that the EU was notified: they then issued a warning that Belgian chicken, eggs and products containing these substances were being subjected to safeguards and restrictions. Days later, pork and beef were added to the list, and then milk and other dairy products.

The crisis led to the resignation of Belgium's ministers of health and agriculture and the Netherland's agriculture minister and his deputy. The premier of Belgium also got the chop, having served longer than any leader in Europe. Regulations were set to be tightened and we thought it was all over, until two years later it happened all over again.

On a very cold day on a farm in southern Holland, I stood listening to Jan Velvaert's appalling account of what had happened to his sows earlier that year; some of the details were so distressing that they stuck in my mind for a long time afterwards. Jan told how he noticed that when his sows were ready to give birth, or farrow, they were not progressing and clearly in distress. While bursting with piglets, their birth canals were not opening and the piglets struggled to leave the mother by any means, resulting in huge suffering, haemorrhage and eventual death. Jan's wife broke down during her descriptions of what she had seen take place in their farrowing sheds. On the day we visited to interview the family, their sheds lay empty, their farm had been closed down and, along with thousands of other animals in Holland, their stock had been slaughtered. Pig farmers who run big intensive farming businesses like Jan's tend to be pretty tough people. As we drove away from their home I couldn't help but be shocked by the anguish of this family. Whether they knew it or not, the trauma of the events they had witnessed was etched all over their faces.

Early in the crisis, it was Jan himself who sought help, insisting that his pigs be tested for chemicals at the slaughterhouse. It soon became evident that their feed had been contaminated with the hormone NPA, which had come from Ireland. Incredulously, we heard how the manufacture of the birth control pill by Wyeth Pharmaceutical in Kildare created waste water and sucrose that somewhere along the way had been sold to the animal feed industry. Again, the rules had been clearly broken. From Ireland, the sucrose entered the feed that Jan's pigs ate, altering their reproductive cycle and causing a food scare that cost Europe over €100 million.

And as if that weren't horrifying enough, in a strange twist to the story, Irish consumers had a narrow escape from the same lethal contraceptive pill-juice. It turned out that some of the exported waste from Wyeth had reached the Netherlands only to return home again destined for an animal food manufacturer in Limerick. Thanks to the vigilance of Val Foley, the director of Premier Molasses, his shipment of feed ingredient from the Netherlands was stopped at Dublin Port. At his insistence, the material was tested, revealing the presence of the hormone NPA, the same thing that had destroyed the sows on Jan Velvaert's farm. Agreeing to be interviewed for *Ear to the Ground*, Val said of those willing to make a fast buck at the cost of human and animal health:

> It would amaze me that they would take that risk. Why I don't know, because it's far too dangerous. I think that on the whole, food is very safe but that doesn't protect us from cowboys. I just hope there's not more of them involved in the animal food industry.

CHICKENS WITH PAGE THREE BREASTS

Unfortunately, there were. Around the same time as the Belgian food scare, Irish consumers were to make the unpleasant discovery that while we were messing around with their pig feed, the Dutch were messing about with chicken fillets destined for our supermarket shelves. In 2002, following a tip-off from the UK authorities, the FSAI looked at chicken fillets being imported into Ireland and found that some had been filled with water to plump up their weight by up to forty per cent. In order to hold this amount of water in the chicken meat, collagen had been inserted into the fillets; collagen that didn't come from chicken at all but from beef and pigmeat. Unlike humans, chickens didn't need breast enlargements to have bigger boobs; a few injections of water and DNA from other species did the job nicely.

Most of this chicken was entering Ireland for the catering sector but was seldom labelled as to where it came from. And if beef DNA in your chicken burger wasn't to your taste, in April of that year we discovered that consignments of chicken that entered Dublin Port from Thailand had tested positive for nitrofurons; cancer-causing substances sometimes used in chicken production in Asia. It seemed that outside the EU, and even sometimes within the EU, food processors were adding various unpleasant concoctions to add weight or longevity to their chicken products. Furthermore, it was almost impossible for the consumer to know this: labelling of contents and country of origin is still an issue. And however lacking in transparency labelling in supermarkets might be, we are completely in the dark about the nationality of chickens used in your average 'chicken and stuffing' garage sandwich or chicken and chips hotel dinner. In the catering sector, country of origin is still a complete mystery to Irish consumers, and this makes it even more difficult to distinguish rogue food operators from good ones.

So what on earth do we do? One thing we've learned from these recurring food safety crises is that while regulations on food safety are plentiful, testing isn't. In the period following the 2008 dioxin contamination, it emerged that feedstuffs at Millstream Recycling in County Carlow, the source of the contamination, had not been tested by a Department of Agriculture official in the past thirteen months. Surely this is too long? Department of Agriculture official Dermot Ryan said that the plant in question was using an 'inappropriate' type of oil not licensed by the Environmental Protection Agency (EPA), and it was this that had caused the contamination. But it emerged that neither the EPA nor the Department of Agriculture had responsibility for checking oil products used to melt the feed mixture during mandatory inspections. The Department of Agriculture also pointed out that the pig feed manufacturer was using raw materials from the 'highly regulated' food retail industry. If it's good enough for humans, it must be good enough for animals. Wasn't the contaminated pork itself not food for humans?

Yes, there are tight regulations on food production across the EU, and yes, the majority of our food is produced within these regulations, but regulations must be accompanied by rigorous testing of the finished product if they are to work. Giving farmers and food processors a giant regulatory pain in the wazoo is all very well, but it's worth absolutely nothing if people cut corners and if, as a result of lack of testing, consumers are at risk of being poisoned. The reality is that wherever humans are involved in food production, there will be those who want to make a few extra quid by bulking something up, adding something or throwing in a few contraceptive pills to help things along. And for those who do get caught, in the rare instances when they are tested, the fines are simply not big enough to stop others doing the same.

In this game of cops and robbers, the Food Safety Authority of Ireland was pretty much the only bobby on the beat in Ireland. It deployed its meagre resources intelligently, and while it was by no means able to offer comprehensive testing of all foodstuffs sold to us, it was an increasingly effective bulwark against substandard or dangerous foods entering the supply chain. It had been set up by the then Minister for Health in the wake of the BSE crisis and had been copied by many of our European partners. In his wisdom at that time, the Minister believed that consumers would be best protected by an agency whose sole responsibility was food safety. The minister in question is now the Taoiseach and has had a change of heart. As a cost-saving exercise, the Food Safety Authority is to be rolled up along with the Office of Tobacco Control and the Irish Medicines Board into a single agency. Fags, food and pills—sure, don't they all go in the same hole? Only time will tell whether this will result in a diminution of the FSAI's powers and ability to enforce standards through testing, but you get the uneasy feeling that when the next food crisis hits and questions are asked as to why proper testing wasn't done, this decision will be the reason why.

It also doesn't help that we have handed over control of most of the food that we eat to large industrial-scale growers, producers and processsors, especially of chickens and pigs. If we are to place blind faith in these businesses, the regulations they produce under and the agencies that monitor them, we are also placing our faith in the hope that the individuals who own and run these businesses have our interests at heart.

When you look at the past ten years of food scares in Europe, let alone a longer period of the industrial manufacture of food or the way in which food is produced in developing countries, it's fairly evident that there will be more serious food contamination events in Europe and in Ireland, let alone across the world. For those of us who like to think about where our food comes from, it may help both your conscience and your health to eat stuff from closer to home. After all, a chicken that came four thousand miles and which still commands a cut-price rate may well have something iffy about it. And while not everything we eat can come from our own doorstep or be under our own control, we can be better armed with information about food companies, food processes and regions of the world that tend to do certain things either well or badly. It's not surprising that the biggest contamination of milk products in recent years arose in China, where there was no history of dairying, husbandry of milking cows or processing of milk products. In Ireland we are good at producing dairy products. We are also pretty good at producing food in general. If anything, buying simply, and buying local, can help both your health and your conscience.

Chapter 16 ∾

THE RISE AND RISE OF
THE SUPERBUG

In February 2009, it seemed that the whole town of La Gloria in Mexico's Perote Valley was getting sick. The caseload of patients presenting with flu-like symptoms grew and grew until local doctors were overwhelmed. When state health workers arrived at the end of March, 1,200 people sought their help—rather a lot in a town of 3,000 people. Maria del Carmen Hernandez wanted to keep five-year-old Edgar at home, but her husband chided her not to be afraid of what 'might or might not happen'. So they continued life as normal while more and more around him were diagnosed with acute respiratory conditions. Then Edgar came home from school with a fever. He complained of a headache so bad that it made his eyes hurt. Two children in the village had died a few days earlier, so Maria took Edgar straight to the doctor, who treated him with antibiotics and he recovered in a few days.

But the doctor also took a swab from inside Edgar's nose and sent it off for testing. The results of that test confirmed Edgar as patient zero in what was to become a global swine flu pandemic. A few weeks later, swine flu had reached Ireland, the Netherlands, Germany, the UK, Switzerland, New Zealand, Canada and the United States. At the time of writing, swine flu has not evolved into a killer with anything like the potential of the 1918 influenza pandemic. In fact, there was probably more danger to public health from the stress created by media hysteria than H1N1 was capable of generating on its own. But just because journalists and news editors got it wrong, we would be foolhardy to dismiss the disastrous potential of diseases transmitted to humans from animals.

Strangely, Edgar's was the only positive test for swine flu in La Gloria. Everybody else had a different strain of the flu, but then, only thirty-five people had been swabbed. Even more peculiarly, Edgar didn't work with pigs, nor did anybody in his family. La Gloria is surrounded by over seventy pig farms. Environmentalists had logged complaints of residents of the Perote Valley about the swarms of flies and overwhelming stench emanating from the farms. It was claimed that crowded conditions in the pigpens were ideal for the spread of any contagion. The people of La Gloria believed that while Edgar was the first person to test positive, others infected at the same time had died but had gone untested. As of now, nobody has figured out how and where the H1N1 virus made the leap from pigs to people, but it indisputably did, and the conditions in which we rear the animals we eat probably had a lot to do with it. In fact, there are many eminent vets and scientists, not given to being alarmist by nature, who believe that we are sowing the seeds of our own destruction in our failure to rear animals humanely.

What a bitter irony that at a time when so much of Western popular culture revolves around celebrity chefs and food gurus telling us how to prepare meals that will salve our consciences and improve our health that food has the potential to become the greatest killer humankind has ever known. There is a presenter in RTÉ who for the sake of diplomacy shall have to remain nameless. In the evening as he sits down to watch television, he plays a little game with himself. How many channels do you have to flick through before you hit a cookery programme? It used to be property programmes and before that gardening programmes, but at the moment he claims that you rarely, if ever, get beyond the terrestrial channels and up into the satellites before you happen upon a vehicle for the talents of Jamie, Gordon, Rachel, Nigella or whoever.

Our presenter complains that on a Sunday you'll get no further up the remote control than RTÉ One and *The Restaurant*. Then after that there's the Kevins (Thornton and Dundon) at each others' throats in some TV cook-off kitchen game. The same on a Monday before Richard Corrigan stops you in your tracks. Then while Corrigan is taking a break, the Ballymaloe stable leap into action. Loosen up the rules a little bit and TV3's predictable offerings like *Top 10 Celebrity Diets of All Time* catch you out. Then it's virtually impossible to get past the BBC stations without tripping up over a *Masterchef*, but if you do, Gordon Ramsay and Hugh Floppy Wellington-Boots are sweeping along the goal line of Channel 4's schedule with what feels like 'twenty-four hours a day of f**king cookery programme broadcasting'.

One suspects that our presenter's irritation might be born out of envy and/or a regrettable lack of skills with food, potted plants and property. Whatever the motivation, the point is, however snidely, well made. If you want to get your mug on the telly these days, it helps to be a chef.

On television, in newspapers and on the radio, chefs command more attention than the literary critics, the political columnists and, Lord help us, proper television presenters themselves. Irish pubs are in some cases now better known for their food than their beer or atmosphere. Cookbooks swamp the tables of bookstores everywhere. Even in cynical Paddyland, we have elevated food, put it on a pedestal and fetishised it. Imagine: there are people who would organise their social calendar around what Trish Deseine is preparing in her Paris kitchen rather than go out for a meal with friends. Between one-fifth and a quarter of the adult television-watching population in Ireland will get stuck into a Rachel Allen TV programme. But as we now spend twice as much on food in restaurants and outside the house as we did twenty years ago,[1] perhaps only very few of us are getting stuck into preparing one of Rachel's recipes.

GASTRO-PORN

Actually *making food* is really not what these programmes are about. At the centre of their appeal is that they are gastro-porn. Nice stuff to look at. The fantasy of actually getting involved in the fray yourself remains frequently just that; a fantasy. Similarly, who really wants to have one of Diarmuid Gavin's concrete monstrosities at the bottom of their garden, or allow Duncan Stewart to start punching holes in their roof? Cookery programmes allow most of us to cling to the comforting myth that we could prepare Coquille St Jacques from scratch because we saw Jamie Oliver teach a room of dinner ladies how to do it in three minutes flat. The idea behind lifestyle programming on television is not that you actually go and live the life: it's more a reflection of the values to which we would subscribe if we weren't so busy watching television.

In Britain, where they have researched the link between gastro-porn on TV and people's eating habits, Martin Caraher from London City University noted, 'If cookery programmes actually had any impact we'd be the healthiest nation on the planet. But we're far from that.'[2]

And what we like to watch is very interesting. Rachel Allen serving up her comfortingly familiar and highly pleasing to the eye formulae can expect an audience of around three hundred thousand people on RTÉ. When Hugh Fearnley-Whittingstall presents his Channel 4 programme challenging our assumptions about how chickens are raised, only about thirty thousand Irish viewers will tune in. Part of that difference can be put down to the stubborn preference of Irish viewers for Irish programmes, but a big part of it has to be about the content. When a programme probes the unpleasant realities of where our food comes from, we are just not interested. That's not what we say publicly, of course. In a survey, eight out of every ten Irish people said they felt it is very important that animals be given access to the outdoors, have a generous amount of space and high-quality food and are slaughtered in the most humane way possible. So does it follow that eighty per cent of Irish people will buy free-range, organic eggs? Unfortunately, only three per cent of us have made that leap.[3]

The same survey by the National Food Centre established that what matters most to us are issues like traceability, antibiotics and contamination, but we don't seem to make a connection between them and the way in which animals are reared. What an astonishing failure of imagination. In the 'farm to fork' equation, the fork is the only bit we appear fussed about. We can't quite join the dots between real-life animals and what ends up on our plates. Yet we are all foodies now. Food has been elevated to the level of art. It has its own social discourse, its own heroes and pin-up boys and girls. Perhaps one of them will fill in the blanks for us by explaining the link between poor animal husbandry and the diseases we end up catching as a consequence.

In the summer of 2008, a friend of ours and her husband spent a month travelling in Greece. The weather was beautiful, the sea very blue and all seemed good in the world. The couple enjoyed pleasant evenings dining on fresh fish cooked at street stall fires, plenty of feta cheese and breakfasts of live yoghurt and fruit. Then one evening, following a chicken meal at an upmarket restaurant, our friend began to feel ill and suffered violent vomiting. What began as a bout of food poisoning quickly turned into a major health event worthy of inclusion in a Sky One special on 'When holidays go bad'. Over two days, our friend became seriously sick. The seriousness of the illness resulted in her suffering a burst appendix, and an ambulance airlift was followed by emergency abdominal surgery in Athens. She was left with a six-inch scar across her belly and a traumatised husband who had thought she was going to die during the four hours of surgery.

The culprit of her terrible experience was salmonella, and it appeared that she wasn't the only one to suffer from it that summer. Hundreds of people across Europe, the UK and Ireland suffered from a serious outbreak. In England, a seventy-seven-year-old woman died from complications after contracting salmonella poisoning. Around the same time, authorities in the United States were tracking the biggest salmonella outbreak since 1985.

Salmonella is a pretty serious bug; in fact, it's a collection of bugs: there are over 2,300 types of bacteria in salmonella's happy family, their place of residence being the intestinal tracts of animals and humans. They are usually transmitted to humans when we eat food that has been contaminated by animal faeces. If you contract salmonella poisoning, your best-case scenario is a bout of diarrhoea and vomiting. The worst-case scenario—death—is usually confined to those with weakened immune systems: older people and pregnant women and their unborn babies. In rare cases, salmonella poisoning can develop into Reiter's syndrome, which is a painful swelling of the joints, irritation of the eyes and painful urination; it can lead to chronic arthritis that is difficult to treat.[4]

In a strange coincidence, production was halted that summer at the Dawn Meats plant in County Kildare in order to carry out a 'pharmaceutical grade cleandown' of the entire plant after suspicions were raised that food coming from the plant might be causing health problems. A statement issued by Dawn Farm Foods said that a link between their plant and the European salmonella outbreak was possible, and associated one specific production line with the outbreak. This production line was closed after the company was contacted by the Food Safety Authority of Ireland. The outbreak prompted the withdrawal of beef strips, chicken, lamb and pork supplied to at least eight European countries.

The complex path of food, how it is farmed, processed and where it travels to, again came under scrutiny, even those foods produced under what we would consider safe, well-regulated conditions in Ireland. This time it was not humans meddling in the food chain that resulted in the contamination, but pathogens raising their ugly heads and causing another crisis of faith in what we eat.

THE NEW KILLERS

Salmonella poisoning, or salmonellosis, though not a newcomer to the scene, is one of an increasing number of diseases that are causing severe anxiety in the world of animal and human health. Salmonellosis is known as a 'zoonotic' disease, something that is transmitted from animals (both wild and domestic) to humans, or even from humans to animals, where they mutate a bit, and then back to humans again. Zoonotics are the focus of increasing attention and alarm. Although they sound far too exotic to be picked up in your local supermarket, zoonotics now make up a whopping sixty per cent of human pathogens and seventy-five per cent of emerging diseases that we are now seeing. Many serious diseases fall into the zoonotic category, including well-known personalities such as brucellosis, leptospirosis (Weil's disease), rabies and, the best known of all, the HIV virus. HIV was originally acquired from primates in Africa, brought into the human world probably through hunting and eating monkeys, or possibly through an animal bite.

Among the new global threats to human health, there is a growing list of emerging zoonotic diseases. Avian flu, West Nile virus, monkeypox and Nipah virus are causing severe anxiety among those involved in forecasting and modelling disease around the world. Zoonotics are efficient and often horrible killers; each year, about 55,000 people worldwide still die from rabies. Rabies is a particularly frightening example of the devastation that can be caused by animal-to-human disease, but much of what we risk contracting from animals is far more mundane, involves no foaming at the mouth but has far wider health implications than the dreaded mad dog disease. It may well come from something we eat. It's as simple as that.

And guess who are the culprits? We are. The WHO identified that it's not animal health or behaviour that has caused this huge threat to our own health, but the ecological impact of our own activities, including farming and food production itself. It's clear that if we don't look after animal health, we ourselves will be the ones to suffer long-term and perhaps fatal consequences. At a conference of European vets in 2008, Bernard Vallat, Director of the World Organisation for Animal Health, identified how 'animal health is a key component of food security, and we need to stress the link between animal health, food safety and public health'. Old policies need to be adapted, he added, in view of the changing world, and it was clear that 'we need to communicate'.[5] He also underlined the importance of the quality of veterinary education, 'which urgently needs updating'.

International travel, global warming, trade in exotic and wild animals, encroachment of humans and domesticated animals into wildlife habitat and concentrated agriculture operations in close proximity to human populations were listed as prime reasons for the rise of emerging disease threats. 'This new era of emerging diseases will continue and maybe even accelerate,' said Dr Lonnie King, the Dean of Michigan State University College of Veterinary Medicine, who also identified that we are 'one plane trip or one import away from a major epidemiologic event'.[6]

What's clear is that the emergence of new zoonotic pathogens seems to be accelerating as transportation has advanced, making it possible to circumnavigate the globe in less than the incubation period of most infectious agents. And because we're all so in love with travel, what emerges in one continent can very quickly end up on an A&E trolley halfway round the world; a monkeypox outbreak in Gambia travelled to the US via the importation of infected Gambian rats and quickly spread to the cute little American prairie dogs, who had never even seen West Africa on the telly. These new diseases aren't picky about their victims; they are fast, versatile and admirably adaptable to the environment we have introduced them to.

UNHEALTHY ANIMALS, UNHEALTHY CONDITIONS

Severe Acute Respiratory Syndrome, or SARS, stemmed from the trade of wild ducks sold as food in high-density marketplaces and in turn affected the humans who had close contact with them. While we may think animal welfare in the food chain is a nice thought if you can afford it, the speed at which SARS spread through its animal and then human hosts was directly linked to the inhumane, packed conditions in which both the Chinese birds and their owners were operating. Bad conditions enhance the spread of disease; sick animals will produce sick humans. It's bizarre that despite the armoury of science at our fingertips, the pressure to produce more and more food as cheaply as possible is raising issues that we should have knocked on the head a hundred years ago when bacteria like salmonella were first identified.

Trade in animals and ecological and environmental changes brought about by human activity are large factors in driving the rise of these newer diseases. So harmful is the potential of zoonotics that they are also the weapon of choice in those interested in bioterrorism. But the biggest contributor to the rise of emerging diseases is a pretty simple one: the upward growth of human and livestock populations across the world, bringing increasingly larger numbers of people and animals into close contact. And as the huge appetites in India and China turn towards animal protein, cattle and chickens will be farmed in ever-increasing numbers in conditions that would not pass muster on this side of the world. You can be sure that where animals are mistreated or reared in extreme factory unit conditions, disease and health issues are ever present. And what ends up inside an animal's system emerges at some stage on somebody's plate.

Antibiotics and treatments for new diseases are redundant before they're even administered. Irish vets are increasingly aware of the problem, particularly the silent changes taking place among the pathogens that threaten both the animal population and ourselves. Peadar O'Scanaill stepped up to tell farmers the unpopular news that:

[T]he bottle of antibiotics (for milking cows) came into being a little over 60 years ago. With underuse, overuse, misuse and abuse, we have created super bugs in the animal world and more worryingly, in the human world. Multiple resistant bacteria are a very big threat to human health at this moment, and we must never, in the food producing world, do anything that adds to that problem.[7]

It's clear that there are warning signs, but what we don't know is how seriously the relevant legislators are taking the warnings. Last year, EU ministers were told that Europe is not prepared to face diseases that may be coming down the line. Dr Albert Osterhaus warned that we need to learn quickly from how incidents such as BSE and SARS moved from animal to human infection. He predicted that avian flu could transform into a human virus and cause widespread epidemics. Is this going to happen? 'Yes', he said. Are we prepared? 'No.'[8]

Vaccination for disease prevention is one option. In 2006, during the bird flu epidemic, Vietnam operated a strict vaccination programme which saved it from the fatalities suffered in its neighbouring countries. But Europe has not reached agreement on how or what to vaccinate against across Europe. The French are currently vaccinating forty million animals against blue tongue disease, but other European ministers disagree on the policy. It's evident that there needs to be more co-ordination across borders and human and animal health disciplines to keep tabs on what horrors may be coming down the line. Biosecurity on farms and at borders cannot be overlooked and it's clear that we already have some problems already, even within Europe. In 2006, testing on pigs across the European Union revealed that salmonella was estimated to be present in one in ten pigs slaughtered for human consumption. The special Task Force of the European Food Safety Authority (EFSA) found the levels of salmonella detected in pigs varied from zero to twenty-nine per cent between member states, and salmonellosis was the second most reported cause of food-borne diseases in humans in Europe; 160,649 people suffered from it in 2006.[9]

At least salmonella is something we can test for. It is also something that consumers can be aware of, and we can push for animal protein that is reared in disease-free environments and conditions of best practice. If we'd rather not know what goes into our mouths, we are in effect putting our health at risk through ignorance. And if we sense that something is too good to be true, or too cheap to be real food, it probably is. Corners cut in animal health mean corners cut in food production. A better working example of the buyer beware principle would be hard to find.

In terms of what vets and disease monitors can legislate for, there are still the unknowns, pathogens that are as of now developing into the epidemics of the future. What can we do about these? In general, there is no way to predict when or where the next important new zoonotic will emerge or what its ultimate importance might be. A pathogen might emerge as the cause of a geographically limited disease, intermittent disease outbreaks or a new epidemic. No one could have predicted the emergence or zoonotic nature of the bovine spongiform encephalopathy prion in cattle in the UK in 1986, and certainly not the species-jumping emergence of HIV as the cause of AIDS in 1981. For the thirty Irish people who have died from CJD since 1980, it's little solace that many of them most likely contracted the disease from the BSE-infected cattle that went, without our knowledge, straight into the food chain.

YES, WE KNEW ABOUT THE CONTAMINATION, WE JUST DIDN'T THINK IT IMPORTANT ENOUGH TO TELL ANYONE

And what about cases where something nasty has escaped the safety loop? Can we trust that food companies are going to act in our best interests and inform us? Dawn Meats's problems seemed to have been addressed in a reasonable amount of time, but we know from the past that some food producers have known of a hazard but failed to act, as was the case with Cadbury, which was forced to recall a million chocolate bars here and in the UK after it emerged that the company knew of suspected salmonella traces in its products. Traces of the rare bug *Salmonella montevideo*, found in some of their chocolate products, was the same strain that resulted in forty-two people in the UK receiving hospital treatment for the contamination. The source of the outbreak was traced to a leaking waste water pipe at Cadbury's Marlbrook factory. And while salmonella was found in some of the firm's products between January and March 2006, Cadbury only began the recall from 23 June. For their sluggish behaviour, Cadbury was fined £1 million and hauled through the courts.

Worse than that is when not just the food companies but their governing authorities start hiding the presence of disease in something that we are eating. In 2007, the *Washington Post* broke a story on how the FDA had known for years about salmonella contamination problems at a Georgia peanut butter plant and on a California spinach farm that led to disease outbreaks that killed three people, sickened hundreds and forced one of the biggest food recalls in US history. It seemed that the FDA was overwhelmed by the growth in American food processors and imports and instead of taking steps to address the problems relied on producers to police themselves. The outbreaks alerted the US to an urgent need to change the way the safety agency does business. Robert E. Brackett, director of the FDA's food safety arm, admitted: 'We have 60,000 to 80,000 facilities that we're responsible for in any given year.' Explosive growth in the number of processors and the amount of imported foods means that manufacturers 'have to build safety into their products rather than us chasing after them,' Brackett said. 'We have to get out of the 1950s paradigm.'[10]

Oh dear. Leaving food producers to regulate themselves didn't work very well, did it? It's clear that with some of these cases, and the really scary ones at that, we only find out about the contamination years after the event when the horse has decisively bolted. In Ireland, the Food Safety Authority has proved pretty brisk in investigating lapses in hygiene and safety standards. In 2007, they investigated eighty cases of food contamination in businesses and food manufacturing plants in Ireland and warned that businesses will not only face hefty fines, but lose customers if it's revealed that their standards are not up to scrutiny. One can only hope that the government's proposals for the FSAI (see Chapter 15) don't undermine this level of inspection.

WOW, THAT BURGER IS GOOD

Aside from the impact humans are having on our own and animal health around the world, there's the stuff that we actively put *into* the food chain to improve some aspect of colour, taste or smell, or to make the poor item of foodstuff last at least a year longer on the shelf than it normally should. Some people call it food processing, others call it food adulteration. Alongside prostitution, it's one of the oldest professions in the book.

Manipulating genetics is nothing new. Ever since humans began cultivating land and breeding livestock, particular strains were favoured over others, leading to the food environment of today, where the animals and vegetables that we select for the dinner plate are those that function the best for our needs. While the idea of genetically modified organisms (GMOs) frighten the life out of many consumers, taking a detailed look at a Belgian Blue bull will probably have the same alarming effect, and this is an animal produced by selective breeding many generations before GMO technologies became available.

The Belgian Blue is the Mr Universe of the bovine world, having a high incidence of muscular hypertrophy; a heritable condition in cattle that primarily results from an increase in the number of muscle fibres relative to normal cattle. Belgian Blues are basically great at converting feed into lean muscle and giving a higher percentage of the most desirable cuts of meat. They have less bone, less fat, and on average twenty per cent more muscle. But such desirable traits often come at a price. Like the pretty Dalmatian dog whose elegant spotted coat accompanies a likelihood of profound deafness, Belgian Blues are not without their problems: they are less tolerant to stress, have more fertility problems and what is kindly called in the business 'calf viability issues'. Think about it.

On a visit to the smallholding of a charming elderly farmer in South Belgium, it was alarming to see how all his cows were stitched, Frankenstein-style, vertically down one side of their body. The farmer explained excitedly in French and with much waving of arms how the calves he now bred were simply too big to pass through the pelvis of his cows. The size of them, so big, so gigantic! *'C'est incroyable!'* All were caesarean births, a costly operation for both our farmer and his cows, but one deemed worthwhile for the stupendous size of their giant offspring. And before we all *tut tut* about modern technology driving our desire for crazy-looking livestock, the occurrence of double muscling in cattle and a wide-eyed excitement about its meat-producing possibilities was first documented in 1807.

A recent internet report on Belgian Blue cattle generated huge comment from readers who thought that the published photos of Belgian Blue cattle were 'gross' and wondered whether 'chowing down on one would give you a six pack'. Strangely, these meat consumer bloggers seemed innocent of the fact that most of what they put in their gobs every day, whether it be a marshmallow or a fish burger, is heavily manipulated in one way or another. The unsavoury fact is that each time we manipulate something organic to exaggerate its desirable aspects, we have to accept that this comes at a price.

Genetically modified foods are a growing reality, but a complex one. On one hand, genetically modified animals and plants present greater production opportunities to meet the world's growing appetite for food. GM crops can be made hardier, disease resistant and produce higher yields. After all, why shouldn't Africans have access to drought-resistant maize crops? Isn't it better than having people starving? And what about Ireland, where the development of a blight-resistant potato is not far off reality? Spraying potatoes weekly for blight is a costly and ultimately environmentally damaging exercise: the spray kills the blight, but it also wipes out a hell of a lot of other micro-organisms which live on the plants and in the soil and are doing no particular wrong to anybody.

On the other hand is the argument that once you start messing around with genetics, there is no going back. Biodiversity in a given area is damaged by cross-pollination with genetically modified crops or plants. Your average Irish wild rye grass might never be the same again after it's had a few years hanging out in fields neighbouring a genetically modified competitor. In fact, it mightn't be able to reproduce itself, and this is one of the main worries about GM cereals in particular. For thousands of years, farmers all over the world collected seeds from one crop to begin next year's cycle. The seeds of many GM crops don't reproduce the same crop again, so farmers are forced to buy seed yearly from the big multinational firms that now control this arena. It's a model that's not good for farmers or good for anyone who likes to dine occasionally; essentially, we are handing over control of food to corporations whose concerns are profit margins and not whether people have enough to eat.

In Ireland we have resisted growing genetically modified crops (except to trial them), but we have to be real about what's going into most of our animal feed, which, particularly in the case of pork, is genetically modified. Pigs apparently go mad for GM soya, it's a good-value foodstuff and makes up the biggest proportion of Irish pig feed. GM cereals are also imported for use in Irish cattle feed: they are cheaper to buy than non-GM cereals, as they produce much greater yields. These days, getting your hands on non-GM cereals for cattle feed is actually a bit of a challenge. In a sense, we are kidding ourselves in Ireland that we are GM free. GM offers great opportunities, but it's clear that it must be handled very carefully in terms of who controls those opportunities and how the technology is rolled out. Intensive farming might feed the world, but it could also irreversibly mess it up.

In contrast to America, Europe has been a little slower to adopt the more extreme types of intensive farming, use of GM crops and vast animal-producing units that are common in the US. Looking at the scale of some American operations, the future looks to be super-sized farms, but producing animals, in particular, that thrive under these conditions is often easier said than done. Animal units unfortunately tend to behave differently from other mass-manufactured items. As opposed to cars, laptops or iPods, animals tend to remain sentient throughout their production process, an irritating tendency that can result in sickness, stress or death of your unit.

A recent report found that that even on an economics-only basis, many realities of modern-day factory farms are pretty stark. On intensive US dairy farm units, ten per cent of cows die each year; twelve per cent have stillborn calves. The animals are stressed, points out geneticist Dr Gary Rogers, and these are high-yielding Holstein cows, animals bred to withstand the system. But even they are being pushed to their limits and will die in numbers that are not really acceptable. What's interesting is that Dr Rogers's research comes not from an animal welfare platform, but an agribusiness one.[11] Animals dying simply because they can't cope with the farming system itself is not good for either farmers or people who like the odd gourmet burger. Animals die because their health fails and, guess what, animals with underlying health issues are not generally good to eat. Ummm.

OH GOD, ANOTHER FOOD LABELLING SCHEME!

At first pass it all seems pretty intractable. Survey after survey has shown that although consumers say they are willing to pay for more humanely and ethically produced products, they invariably don't follow through and actually buy them. Irish consumers in particular don't see themselves as a part of the equation. We feel that responsibility for animal welfare lies with welfare organisations, vets and farmers 'rather than with consumers'.[12] So what do you do? How do we encourage ourselves to use our purchasing power to re-shape the market for our own benefit? It's a tough one because in spite of our professed concerns about health, animal disease and welfare, and contamination, our two major priorities when shopping are consistently convenience and price.

There might be a way out of this conundrum, though. When surveyed, Irish consumers have indicated they feel that very often they are shopping in ignorance and would like somebody to step in and lead the way. We have shown significant support for the idea of an EU-wide label that would set out the animal husbandry standards of the chop or steak you are about to purchase. Another labelling scheme? Surely not! Perhaps those questioned were getting hacked off with the person with the clipboard and were prepared to acquiesce to whatever lunacy they suggested if it made the questioning stop. We don't like labelling schemes; there is buckets of evidence to show our resistance to them. But it is only right and proper that consumers throw this responsibility back onto the market and government. After all, no shopper can individually interrogate the origins of every single meat or dairy purchase they make. It is government's job, not ours, to examine, explain, enforce and regulate. From the shopper's point of view, such a scheme need be no more

complicated than bronze, silver and gold yardsticks on the product with a
website address providing more detailed information if you are in the mood.

Take, for example, poultry, which by 2012 will have a whole new set of
standards when battery production is outlawed. This doesn't mean
production standards will be uniform, though. There will be 'free range', 'barn
reared' and 'enriched' cages in which the birds will have a nest, litter, perch
and clawing board. Within these three approaches there will be those who
favour de-beaking, not providing natural light, organic feed, accelerated
rearing processes, and so on and so on. The eggs or chicken fillets produced
by these birds can be marketed in any number of misleading ways. Wouldn't
it be so much better for all of us if the bureaucrats told the industry what they
would have to achieve in order to be certified? They live for that kind of thing
in Brussels.

You can expect squeals of opposition from the producers and farmers at
the prospect of even more bureaucracy, but if it were married to enhanced
subsidy for those meeting the higher standard, they would probably be
mollified easily enough. So the producer who individually plumps the pillows
of his broilers and reads them a bed-time story will be able to bring his
product to market at competitive prices. The mucky stuff coming in from
South-East Asia or Brazil will get no certification, but at least consumers will
know they are paying their money and taking their chances. No doubt new
forms of state support of any kind for farmers will prove unpalatable to some.
Others will baulk at the idea of more 'nanny statism'—the people in Brussels
who brought you the straight banana now decree how many square feet
constitutes freedom for a chick to roam. Surely, though, by this stage we all
have realised that the commercial markets don't regulate themselves, and
consumers don't have the time or inclination to act as food detectives.

Change will ultimately be motivated by consumers and how much they
care about what they eat, but it will have to be led by policymakers. It would
be nice to think that we could all establish relationships with our butcher or
grocer and know that the beef or chicken we are eating came from decent
conditions fifty miles down the road. That by shopping as locally to home as
possible our wallets could become agents of change. It works for some, but for
the rest of us there are just far too many good cookery programmes on the
box to have the time for that kind of thing.

Chapter 17 ∾

THEY DON'T GO AWAY, YOU KNOW

Philip unearths the perennial problem with pesticides

We were well into the ham sandwiches and on to our second pot of tea when the ladies dropped their bombshell. A preserved example of another time and place, Mary Flynn's kitchen was of the kind that now appears in glossy coffee table books dedicated to chronicling the passing of another era. Linoleum floor covering, Aga with a permanently simmering kettle, dog's basket in the corner. In front of us sat the big thick slabs of ham pressed between thick buttered doorsteps of old Mr Brennan's best. And mountains of them. You'd break your wrist trying to lift the plate with one hand.

But as we sat around the table listening to the mushroom pickers' story unfold, we realised it was a tale with an entirely twenty-first-century dilemma. Myself and my colleague, the brilliant RTÉ radio producer Alan Torney, had been nosing about in the mushroom industry for a couple of months. The pickers were alleging exploitation by their bosses in what had become a hot media topic. But like so much public debate, it was informed by anonymous allegation and generalised accusation. At this point we were trying to persuade workers to come out from behind their cloak of anonymity and make their claims publicly. It wouldn't be so easy for the industry bosses to dismiss them then.

Weeks of phone bashing had led us to Mary's door just outside Ballaghaderreen in Roscommon. She was a dream interviewee, who had meticulously held onto every payslip she had ever received from her employers, Shannonside Mushrooms. 'I'm a hoarder,' she chuckled. The slips were documentary proof that workers were being paid as little as €3.50 an hour when the agricultural minimum at the time was well over €8. Startling as her claims were, stories often take the most unexpected turn and sometimes the topic that you start out investigating becomes less of an issue as you dig deeper. So it was when Mary and her former co-worker, Elizabeth, started telling us about conditions inside the plastic mushroom tunnels. As myself and Alan listened with increasingly unsettled stomachs, there wasn't another sandwich touched before we left Mary's house well after midnight.

The ladies claimed that their skin had become cracked and sore, their hands had bled and their breathing was impaired after exposure to chemicals which had been sprayed on the immature mushrooms. Most days they would be sent into the tunnels to pick within hours of the spraying. Rubbing her fingers gingerly, Mary recalled how 'the skin on our hands would be kind of scalded and we had difficulty breathing. Like you had laryngitis all the time.' While some of the symptoms differed from person to person, there were common denominators. All had increased instances of upper respiratory tract infections, streaming eyes, damaged skin and a burning sensation in their nasal passages. Any protective gear they were offered had to be bought from the company, and Mary claims that if they ever complained they would be shown the direction of the gate and told that if they didn't like it they could 'go f**k off and lay blocks' for a living instead.

Throughout the interviews, Alan kept on giving me pointed looks. He didn't have to say it out loud; we both knew that the next logical question was very worrying. If this is what the chemicals were doing to the pickers, what might the effect be on those who were consuming them? We continued to poke around Ballaghaderreen, but we were hamstrung by the fact that RTÉ would not tolerate us trespassing on Shannonside Mushroom property in pursuit of the story, so it was difficult to establish exactly what chemicals and in what quantity were or weren't being sprayed on the crop.

Eventually Alan found an employee who was willing to take photographs of the drums of chemicals. The photos wouldn't win any awards, but they clearly identified three different substances, two of which had no reason being anywhere near mushrooms. A bit more digging unearthed a former employee who claimed to have actually been instructed by one of the company's owners to do the spraying. Back home in his native Bulgaria, Arcadie Cretu spilled the beans on how one of Shannonside Mushroom's owners had instructed him when he was a plant supervisor to spray the produce with a cocktail of chemicals so that they would be entirely blemish-free. The air in the tunnels was 'very bad. I feel headache and sometimes vomiting. I was always under depression,' he claimed.

The purpose of the exercise was, ironically, to remove unwanted fungi from the fungus. Most consumers shop with their eyes rather than their taste buds and button mushrooms that were anything less than a uniformly pristine white tended to get left on the shelf, if not rejected by the retailer altogether. Toxicologists we spoke to were alarmed, not least because one of the chemicals, formaldehyde, was a known cancer-causing agent. The other two that were sprayed may also have been carcinogenic, but knowledge about their effects on humans was limited because, guess what, nobody ever really tests them on people. We told the Department of Agriculture what we had found, yet somehow—without testing a single mushroom from the plant— they were able to assure us that there was absolutely no risk to public health. Why? Well, because when they called to the mushroom plant they didn't find any of the chemicals on the premises, and on that basis, why bother testing for them?

The Hogan brothers were not hugely impressed by our investigation. They never denied any of the allegations about what they were spraying on fourteen per cent of the mushrooms produced in Ireland, but they did go on the offensive. I ended up helping Gardaí with their inquiries, having been accused by the brothers, under some arcane piece of nineteenth-century legislation, of 'trapping, watching, intimidating and besetting' them. I have yet to be charged with anything. Not long afterwards, Shannonside Mushrooms closed down and they were involuntarily struck off the Companies Register.

So what happened next? Did somebody conduct laboratory analysis of their mushrooms to assess what risk, if any, there was to public health? An investigation by Bord Bia of how this company could have won quality assurance awards? Inquiries from the Department of Finance as to who had decided that this company was a worthy recipient of hundreds of thousands of taxpayers' euros? Not a bit of it. The Food Safety Authority said it had nothing to with them. Bord Bia had no comment to make, and the Department of Agriculture poured scorn on the whole story. Only one taxpayer-funded quango was moved to take any action. And fair play to the Health and Safety Authority: prompted by fears for worker rather than consumer safety, they made unannounced inspections of mushroom farms all over the country. They found the same chemicals, which had as much business being on a mushroom farm as the Bolshoi Ballet, present at a 'significant' number of mushroom-growing facilities.

BEAUTY QUEEN MUSHROOMS

Clearly Shannonside Mushrooms were not the only mushroom growers who were dosing their produce with these toxic, caustic and, in sufficient quantities, carcinogenic chemicals. Not a single other government agency was moved to take any action. How, then, you may ask yourself, could something like this pass without detection? Why is it that testing didn't show up any pesticide or chemical residues in any mushrooms throughout this period? Like little monkeys covering their eyes, mouths and ears, it seems that if you don't ask the evil question, you won't hear any uncomfortable truths. We produce seventy million kilos of mushrooms every year and official results suggest that they are all absolutely hunky dory. Why wouldn't they be? We only test eleven mushrooms a year. No, not eleven tons. Not eleven punnets. Eleven *individual* mushrooms are selected each year as a statistically representative sample of every other mushroom in the country.[1]

As many as 7,000 million individual mushrooms are grown in Ireland every year. If, as is the case, only eleven special examples, on whom the honour of representing their entire species falls, are tested, the resulting ratio would amount to a percentage that my calculator doesn't have enough screen space to display. Dropped needles in oceans of mixed metaphors and haystacks doesn't even begin to express how utterly insignificant these tests are. A career guidance teacher telling pupils to ditch the Leaving Cert and do the Lotto every Wednesday would be giving his pupils a statistically better crack at a steady income. But perhaps I'm being unfair to the department? Perhaps the testing regime is targeted at the bad boys? Maybe a crack team of white-coated lab technicians, armed with test tubes and mass spectrometers, swoop unannounced in the dead of night on mushroom tunnels? No, unfortunately not. Well, then, surely there must be some kind of random selection system picking on mushroom farmers who know neither the hour nor the day of their choosing and so will be encouraged to behave all year round? Nope, wrong again. The department's routine analysis doesn't involve going anywhere near a farm. They get on to a wholesaler and ask him nicely to supply them with a mushroom of his choosing for analysis. Such conduct is all very civilised, but would you really continue to fight under the Queensberry rules when your opponent steps into the ring carrying a baseball bat? The Department of Agriculture's justification unwittingly acknowledged how pathetic the testing regime was. They didn't attempt to argue that a handful of mushrooms was a reliable sample, but merely that 'they only test fourteen in the Netherlands'.[2] 'Sir, Sir, yes I did pull Jacinta's pigtails but Seamus kicked her in the shins.'

In 2006, of the eleven little button cups plucked from fungal obscurity and pushed under the glare of a mass spectrometer, four were found to contain residues of pesticides. All levels were below the maximum permitted amount, but follow the logic through to its conclusion. If the eleven mushrooms are representative, then a quarter of all mushrooms grown in Ireland contain traces of a pesticide called Prochloraz.

Of even more concern was the twenty-five per cent of home-grown lettuces that exceeded the maximum permissible levels of lovely-sounding concoctions like Azoxystrobin, cypermethrin and dimethoate. And if the one sample of domestically produced spinach that was tested in 2006 contained twenty-five times the permissible limit of carbendazim, could this mean that up to one hundred per cent of Irish spinach is contaminated with pesticides? If you are genuinely in the business of trying to keep contaminated produce out of the food chain, testing eleven mushrooms, four cabbages, three onions and one lonely old leek each year is a bit like catching shrimp with a rod and line and not a bloody great net. But wait, it gets worse. Of the paltry amount of fruit and vegetables that were sampled, over seventy different pesticides were detected, but there were twenty that had no maximum safe level set. So the mandarins imported from Spain and Peru could have been oozing cyprodinil and Propargite out of every pore, or the cauliflower we grew at home might have turned purple with Tebuconazole, and the department would have been clueless as to whether this was a good or a bad thing.

Before you think you can buy your way out of the problem by opting for organics, you should know that sixteen per cent of the fruit and veg that were marketed as being totally chemical-free turned out to contain pesticides. It is thought that they came into contact with the chemicals during shipping from places like Spain, but we don't really know; something that must really annoy the majority of very conscientious Irish organic growers. While you may accept the eagerly offered official verdict that the levels found 'posed no risks to consumers', you might also ask yourself why they were there at all.

WELL, YOU WOULDN'T WANT TO START FROM HERE

It's a bit of a Gordian knot, really. The harder you pull at it in an attempt to untie it, the tighter the pesticides knot becomes. The use of pesticides alongside synthetic fertilizers has been responsible for an unimaginable boom in agricultural productivity. At the end of the Second World War we were getting on average a quarter of the yield per acre that we do today. Chemical innovation born of a desire never to go hungry again introduced a revolution to farming. An ever-evolving range of pesticides, herbicides, insecticides and fungicides are at the heart of a chemical formula which generates more food more cheaply than at any other point in human civilisation. Yet paradoxically, they have now started to undermine our ability to grow. Just as pathogens become resistant to antibiotics, insects, fungi and weeds develop tolerances to the strongest of the chemicals they are doused with. Chemical companies have to develop newer, stronger replacements. For the farmer, first-time use becomes a gateway drug to getting hooked on more and more toxic sprays and potions. Encouraged by government officials and chemical company reps to embrace pesticides, the orthodox view was there was nothing wrong in this. By now it is too late for the food producer to go cold turkey and cut their ties with their supplier. Giving up organophosphates would cost cereal farmers ten per cent of their gross profit,[3] and in an industry with notoriously tight margins this kind of loss is unsustainable. Farmers aren't lying when they say that modern farming would be impossible without pesticides. The tricky part of the knot is that it also may become impossible *with* pesticides. Since the 1960s, crop failures have increased by twenty per cent even where the most up-to-date products were being applied.[4] We are also destroying the very thing that makes growth possible in the first place: mud.

Perhaps soil is the greatest story never told, but it just can't get its public relations in order. The hidden depths of the oceans and the rainforests throw up exciting new creatures or previously uncontacted tribes every once in a while to ensure continued favourable press coverage. That tragic image which David Attenborough and the BBC Natural History Team captured of a polar bear swimming to certain death did more for the global warming campaign than Al Gore and a million pie charts ever could. But mud just doesn't excite the same passions. It stubbornly resists being shoehorned into a romantic narrative of the kind that lends itself to Julia Roberts in the lead role armed with sample jars and a burning desire to expose wrongdoing. But if mud got itself a good agent, a few friendly faces in the media and the Coen brothers to develop a script, it would have a pretty compelling story to tell.

Consider for a moment that a single teaspoonful of healthy mud may contain up to one billion individual forms of life drawn from as many as ten thousand different species.[5] If they were the size of humans, some would clear whole counties in a single bound; others would devour entire herds of cattle in one sitting. Their relationships with each other are every bit as complex as any soap opera, the levels of violence between some species would make Dublin and Meath footballers blanch, and the complexity of the jobs they do is only barely guessed at. There are bacteria that break down molecules using specific enzymes at different levels of the soil in a way that is most advantageous to the molecules being taken up by plants again. There are fungi that wrap themselves around the root system and pass minerals up from the soil into the plant because Mother Nature forgot that bit when she was designing roots. There are armies of mites, protozoans and nematodes that have kept the soil rich and fertile for generations of farmers, but scientists reckon that they have only identified about five per cent of the life forms at work beneath our feet. While we are astonishingly ignorant about this microscopic universe, there is agreement on one thing: blasting crops and soil with chemicals kills the good guys as well as the bad.

For millions of years these guys did their thing, reprocessing dead material and feeding plants while feeding themselves. In the last sixty years we have launched a shock and awe offensive against individual species that interfered with our plans. Any schoolchild will tell you that if you take one species out of an ecosystem, there will be serious knock-on consequences up and down the food chain. Somehow we have failed to apply this universally accepted wisdom to the soil. Perhaps that is because at first glance we appear to have gotten away with it. If the weather allows, tillage farmers can still produce bumper crops. But many farmers making the switch to organic production would believe it's catching up with us. On many farms, the soil is increasingly becoming a sterile medium. It will continue to produce crops but only because it is saturated with synthetic fertilizers that make growth possible. Farmers are increasingly aware that the structure of the soil itself is changing. The good stuff is rich in organic matter and breaks up easily into well-formed crumbs. The not-so-good stuff is dry and sandy or has the waterlogged consistency of potter's clay.

Just like prospectors drilling for oil, farmers now have to buy bigger and bigger machinery enabling them to plough deeper to uncover that rich loamy soil they need. Smart farmers know that damaged soil structure means poorer growth, and that hits them where it hurts most, their pockets. But like the fella offering directions who says 'you wouldn't want to start from here', we are really in the wrong place to make the transition to pesticide-free agriculture. Years of bullying by the food companies and the supermarkets they supply to reduce their costs and decades of official encouragement to use pesticides means that most farmers haven't the financial safety net they would need to leave land to lie fallow and recover. Bankrupted immediately if they do, bankrupted more slowly if they don't. Who'd want to be a farmer?

THEY NEVER GO AWAY, YOU KNOW

Just because 150 of 350 commonly used pesticides can cause cancer[6] doesn't mean that they will. It's more likely that if you've got cancer you should be pointing the finger of blame at cigarettes, your genes, an unhealthy diet or overexposure to the sun rather than at modern farming practices. Even if you have been exposed over the long term to pesticides, herbicides, fungicides and insecticides, the body has a wide range of self-defence mechanisms. Every few days it strips layers of cells from the mouth, throat, stomach, colon and skin as a means of flushing out alien chemicals. Inhaling a lungful of DDT or ingesting a few micrograms of lindane isn't a nice thought, but neither is it a certain death sentence. What caused the cancer will usually depend on whether you are looking at it from the point of view of the patient, doctor, medical researcher or chemist who made the product. The person who can prove with absolute certainty that there is a link between pesticides and cancer will probably have a Nobel Prize medal somewhere about their person, and as the Swedish Committee hasn't made that phone call yet, all such claims should be treated with a degree of scepticism.

Yet there is a nagging amount of research that continues to suggest a link. The US Lymphoma Research Foundation did a very useful survey of up to one hundred major cause and effect studies. Seventy-five of them indicate a connection between exposure to pesticides and lymphomas. Twenty-four show no relationship.[7] So on balance of probability, *doesn't this prove that exposure to pesticides causes cancer?* If only life were so simple. Science has an irritating habit of never proving anything beyond all doubt. There is always room for a researcher employed by the pesticides industry to argue that a disease could be partly caused by some factor not taken into consideration. And they would be right. Maybe there is something that hasn't been thought of at all, a Rumsfeldian 'unknown unknown'. Where chemicals and humans and ecosystems are concerned, the complexity is enormous, the tools of science are crude, and what is not known is always much larger than what is known. Science will never provide definitive and uncontradictable evidence linking smoking to lung cancer or human activity to global warming. But we aren't sitting around on melting ice floes passing round the Silk Cut Purple any longer. There comes a point where common sense dictates (and most scientists agree) that you apply the precautionary principle. Until such time as you know more, you urge consumers to apply a bit of caution.

'So, then, as long as we consume foods with residues below the maximum residue level (MRL) we'll be okay?' You'd think that, wouldn't you? You'd think that an MRL was a level set on the basis of how much of any chemical the body could be exposed to over its lifetime on a daily basis. That would be the logical thing that any sane person would infer from the use of the words 'maximum' and 'level'. Only it's not as straightforward as that.

When you look a bit more closely into what exactly MRLs are, you find that they are actually 'based on good agricultural practice data'[8]. That means that an MRL is the maximum amount of a pesticide that would be expected in a food if it had been applied correctly to a crop. Not what should be going on inside your body, but what should ideally happen in the field. An MRL actually only assures us that when spraying his pesticides, the farmer didn't drive around one corner of a field doing hand brake turns. Very often, in fact, MRLs are set above what the EU should know to be a safe daily intake.[9] Work is being done in Brussels to make a bit more sense of these levels, but there are powerful lobby groups, funded by the agrichemical sector, arguing vigorously for maintenance of the status quo. In a time of escalating food prices, they claim, doing anything that will diminish farmers' output couldn't possibly be in our interests. Perhaps they have a point, but perhaps the continued use of pesticides benefits them more than most.

All medical students are taught, 'first do no harm'. The ancient Greeks had a better handle on snappy aphorisms than the European Commission, but in 2000 the Eurocrats gave modern expression to this precautionary principle. In essence, they said that you just don't do something if 'preliminary scientific evaluation indicates that there are reasonable grounds for concern that the potentially dangerous effects on the environment, human, animal or plant health may be inconsistent with the high level of protection chosen by the EU'.[10] Lofty sentiments indeed, but the reality on the ground is that the pesticide manufacturers sell their product without first proving conclusively that they are doing no harm (remember, seventy-five out of ninety-nine studies linked pesticides to human cancers). The burden of proof of the levels at which pesticide residues can be safely consumed rests with the authorities, not with the companies that make billions from the sale of pesticides. To be fair, you would have to be pretty trusting to take the manufacturers' word on what is and isn't safe. The worrying thing, though, is that the pattern to date is that the authorities are playing catch-up. Take the example of DDT, which was widely used from the 1950s until it was banned in the 1970s. Not to worry, though, because the industry had a replacement waiting in the wings: Lindane became everybody's insecticide of choice. Only after more testing and analysis of what it was doing to human and environmental health, the authorities decided that it too should go the way of DDT. When Lindane was banned in the late nineties, there were others waiting to take up the cause of war against insects. But even as the insecticide Endosulfan was taking off in popularity, fears were being expressed about its safety. Eventually Endosulfan also went the way of DDT and Lindane and at the time of writing the EU has listed dozens of other pesticides for the high jump. But they haven't gone away, you know. They are persistent little fellas and in spite of being banned, they just keep cropping up in the food chain. The most recent figures published by the Department of Agriculture show that more than thirty years after we thought we had seen the last of DDT, it was found in trace levels in Irish pork meat in 2006.[11] Lindane too.

The agrichemical industry now argues that the EU is banning active chemical ingredients faster than it can come up with them. They say that insects, fungi and bacteria are evolving ever faster and crop protection is being put at risk. They could have a point, but it is somewhat undermined by the fact that the organics sector seems to be dealing reasonably well with the same threats. It would be churlish not to admit that while we have reaped enormous benefits from the increased agricultural productivity some pesticides have allowed for, we have also paid a heavy price for others. And the unfortunate reality is that for all our learning, we still don't know enough. For instance, while a certain amount can be said about the impact on human health of an individual substance, little or nothing is known about what happens when all of those chemicals get together and party in our fat cells. We simply don't know whether or not a group of pesticides in our body could be much more toxic than the sum of their parts. The history of these things has shown that we have an unfortunate habit of not discovering the full consequences of a new technology until it has gone wrong.

There is a solution, though; all it involves is actually sticking to that lofty precautionary principle. You shift the burden of proof to those who have created the risk in the first place. Companies exist to grow and serve their shareholders, scientists are employed by these companies to innovate and create new products. These are their first principles and ones to which they remain true. So every time a scientist has a Eureka moment in a lab and synthesises the next big thing, we should say, 'Great, but is it safe?' If there is even the slightest element of risk—and it would be a very foolhardy scientist who would commit him or herself to saying their product is completely safe in all circumstances—we should be allowed to say 'thanks, but no thanks'. It is, after all, what we eat.

Chapter 18 ‿

COULD FOOD AND
FARMING SAVE US ALL?

*Over the past fifteen years the Irish economy has been
transformed from a country of sluggish economic
growth, high unemployment and serious fiscal
imbalances, into an economy which today has the
lowest unemployment rate and one of the strongest
fiscal situations in the European Union. It also now has
a record number of people in employment.*[1]

How like a different country the past seems now. If Ireland was ever to
get the Disney treatment, the Mouseketeers would set it in 2006.
2006 The Movie would be an adoring animated romp of takeaway
coffee cups doing the dance of the teacups and Brown Thomas paper
shopping bags flying in formation through the air over a country of shiny,
happy, wealthy Irelanders.

But just like a Disney movie, at a critical plot point, the sky darkens
ominously. The Irelanders look up in horror as an approaching threat bears
down on their shiny CGI-enhanced world. Oh no! Our high-rise apartment
buildings and chip and pin habit might not prove enough to keep the
economy afloat! One of the major features of our stupendous growth was the
mushrooming of the cost of doing business in Ireland due to inflation, wage
growth, cost of services, property, fuel and adverse exchange rates. Suddenly
we were an economy that stopped growing, we had reached the top of the
Keynesian economics black slope and, whether we liked it or not, we were
putting on our skis and hurtling fairly rapidly back down again. 2006 saw
property prices peak, with 20.2 per cent of jobs created in 2005 and 2006
coming from the construction sector alone. Ninety-three thousand houses
were built that year. In true ostrich style, few of us wondered who was going
to live in them: after all, most of the Poles were already on their way home to
become their own property barons. Let's hope for their sake that they learned
some lessons from us.

SUZANNE FALLS ASLEEP AT DUBLIN CASTLE

The funny thing is that during the good times in Ireland, the MTV Cribs lifestyle that we supposedly lived was not available to all. While incomes rose across the country, the boom didn't touch every home; it was Dublin that experienced the biggest rise in disposable income, with the south-east and fancifully named BMW regions sitting at opposite ends of the chart.[2] In 2002, I sat in a very nice room in Dublin Castle listening in total confusion to various politicians who pranced about in front of a giant PowerPoint display, effusing unintelligibly at the launch of Ireland's long-awaited National Spatial Strategy.

Confusion and boredom are familiar states at government press launches, but this one set the bar particularly high. To a packed audience of journalists, various politicians and strategists talked impenetrably about how hubs, gateways and a lot of Venn diagrams lifted from an Inter Cert maths textbook would solve Ireland's unfair distribution of jobs, infrastructure and wealth. To be fair, there was a germ of a good idea here; Dublin's runaway growth needed to be balanced by industrial growth in another part of the country. But which part should get the goodies? And this is where it got ridiculous. A room full of dispassionate civil servants left to their own devices might have been able to come up with the best solution. Unfortunately, it was a room full of selfish politicians who made the decision, and they all wanted an equal slice of the pie for their constituency. By the time they were finished, there wasn't a townland or a parish pump anywhere in the country that hadn't been bestowed with some kind of special hub or gateway status. Everywhere got a mention. Everywhere would grow. Nowhere would be left out, not even Dublin. Which was especially confusing, as transferring growth away from Dublin was surely the point of the whole exercise in the first place. It was baffling and outrageous and even those giving the presentation seemed to be at a loss as to what it all was really about. Hubs, links, branches … gateways … emm … hubs.

A few hours later, I left Dublin Castle in a buoyant mood; a good dose of comedy in one's morning is no bad thing. The only problem here was that most of the strategy itself was in fact comedy. Needless to say, much of it has been abandoned or was never intended to be implemented in the first place. But at the heart of its failure was not the state's shrinking coffers, but the fantasy and endless optimism we had placed in Ireland's economic growth. The growth envisaged by the National Spatial Strategy was all industrial and construction related. It all relied very heavily on estimates of property values on paper. And things that weren't even built. And some retired grandmothers' savings from Hamburg and some Alabaman twenty-five-year-old Ninja's (No Income, No Job, No Assets) bad mortgage. No wonder it turned out such a mess. Where was agriculture in the thinking of the government? Surely a spatial strategy should have a bit about the single biggest use of space in the country? Or was Ireland, in the government's thinking, now officially a post-agricultural country?

THE FAT AUNT

Unsurprisingly, the National Spatial Strategy didn't succeed in achieving its goal of more balanced regional development. Growth and prosperity, where it happened, are still heavily concentrated in certain parts of the country, particularly Dublin. During the good years, the east coast of Ireland expanded into a Los Angeles-type sprawl of congested road networks and badly planned development, with herds of mini-Paris Hiltons hitting malls with mummy's MasterCard. At the same time, many rural areas were being denuded of young people. Government focus at this time was as far away from farming as it could possibly be. Its gaze was solely focused on attracting foreign direct investment and regulating the growing economy. In this context, rural Ireland and farming activity were less and less important. It was the fat aunt sitting in the corner at the wedding while everyone fluffed and fussed around the spray-tanned young urban bride. Ireland's countryside, and particularly farming, was seen as stuck in the past, lacking the potential to generate wealth, to drive innovation or to keep up with the huge growth that other sectors were bringing to the party. As former Agriculture Minister, Ivan Yates, says, 'We farmed the subsidies for income rather than developing new markets and products.'

Rural Ireland only hit the headlines when rows erupted over planning permission for holiday homes or environmental degradation. After all, for many urban people the countryside was 'holiday land'; a location for luxury weekends, walking holidays and hen parties. In terms of actual policy on rural Ireland, most of it seemed to be defined either by the EU or CAP and our great leaders were happy to keep it that way. It seemed to many that nobody was driving the rural bus any more, or even interested if the bus itself was about to wobble dangerously off the road. There was a feeling that if farming activities were not profiting Ireland exponentially in the way that the building and service sectors were, forget about it. If it's not making us real money, who gives a shite? 'Farmers are moany bastards anyway. If they all died out tomorrow we'd be better off—turn the whole thing into one giant golf club.'

Now that the property boom proved to be the one-night stand that never called us back, it's the agri-food sector (a posh way of saying farming and food businesses) that is held up by some as the best example of an indigenous industry which has a key role to play in our future development. After all, one sure thing about people's economic behaviour is that we all have to eat and spend money on food. Happily, Ireland is pretty good at producing the stuff. In 2007, a report commissioned by Agri Aware totted up the not insubstantial source that farming provides in terms of direct and indirect employment in rural Ireland. Their chief good news could be summed up as follows.

- In the second quarter of 2007, there were 109,700 people employed in agriculture, equivalent to 5.2 per cent of total employment.
- In 2006, farmers purchased inputs and services worth €3.6 billion and made capital investments worth €689 million. Most of the amount spent on inputs was local.
- Farmers earned an income of €2.4 billion in 2006 and most of this was spent in the local economy.
- Farmers produced the bulk of the product for the valuable food and drink industry, which makes a major contribution to Ireland's export earnings.
- Farming activity makes a significant contribution to the economic and social life of rural Ireland. The future of rural communities will be sustained by farming activities.

The survey calculated that agriculture and the food and drink industry accounted directly for 164,700 jobs, or 7.9 per cent of total employment in the economy. If other jobs dependent on agriculture and the agri-food sector are included (areas such as retail, construction, transport, distribution, engineering, further processing and other services), the overall figure was about 230,000 jobs, or 11.1 per cent of total Irish employment.

It certainly makes a compelling argument for paying more attention to what farming can offer us. In a situation where employment in the building sector and manufacturing is falling drastically, and China, India and Eastern European competitors are attracting industries that formerly might have come to Ireland (before we blew the cost of living out of the water), it's clear that the fat aunt in the corner might just provide some cushioned support for our indigenous base. After what could be seen as a period of neglect, farming, combined with a good dose of innovation and the development of our export markets, might provide some obvious employment and economic solutions.

GREEN SHEEN

By accident of nature, Ireland produces really good grass, we have generally good soils and plenty of rain. Farming in Ireland, while it isn't easy, is far easier than it is in much of the rest of the world. Our dairy and beef animals do well living on our natural resources alone for most of the year. In an economy that is now facing so many challenges, it is clear that the fat aunt has much greater potential and can become a more significant component of the Irish economy in the medium to long term. Already at both the primary and production level, food and farming has been a very good servant. It's a pity that something that does so much dog work and adds so much to the economy doesn't get more respect and recognition.

Awareness of our natural resources may provide another key to farming's future success. While sales of organic produce to Irish consumers has been slow enough to take off, it seems Europe is mad for the stuff. If we can convert more farms into organic systems to exploit these markets, Ireland could have a lot to gain. 'It is better to be part of a growing market than a shrinking one' was the message from Professor Ulrich Hamm, Head of the Department for Agricultural and Food Marketing, University of Kassel, speaking in Waterford in 2008.[3] He talked about how major markets such as Germany have seen massive growth in the demand for organic produce, which supply has not been able to match. In organic beef alone, even with a thirty-two per cent rise in the number of organic cattle killed in the past three years, apparently we are still not scratching the surface of export potential for Britain. And with over seventy per cent of organic fruit and vegetables bought in Ireland imported, it's obvious that demand is there.

Interestingly, no other type of farming engenders either the hyperbole or the slamming that organics does. Despised initially as pagan nonsense spread by crazed Europeans living in West Cork, the cult slowly seeped across the rest of the country and, more importantly, into the mouths of Irish chefs and food writers. After a lot of suspicion, organics soon became first a known term and then an acceptable one sometime in the 1990s. And over time there have been big Irish success stories in the sector such as Glenisk, the organic brand of milk and yoghurts run by the Cleary family in Laois. We first visited their farm for an *Ear to the Ground* programme in 1999, when most farmers we interviewed thought organics was a byword for mad people who wanted to opt out of life and dance around tepees. Viewed by many with a fair dose of suspicion, the Clearys turned Glenisk into one of the most successful organic businesses in Ireland, opening a new state-of-the-art processing plant in 2008,

creating thirty new jobs and bringing the total employed in the Glenisk
enterprise to fifty-five. In a climate of rising unemployment across the
country, Glenisk's success is a real good news story and could be a key to more
opportunities out there in the farming and food sector that are yet to be
exploited.

Though some farmers have entered organics only to leave it again, the
experience of other operations that started off on an even smaller scale than
Glenisk has sometimes yielded more than encouraging results. Padraig and
Una Fahy from Beechlawn Organic Farm in Ballinasloe have built a thriving
business on thirteen acres. Padraig gave up his teaching job in 2004 and Una
joined him full time in the business two years later. Over ninety per cent of
their produce is sold direct to the public through farmers' markets and box
schemes. Unsurprisingly, the supermarkets are tricky outlets, as Padraig
observes: 'Many supermarkets demand perfect veg but only pay when you are
on your hands and knees.'[4] But supermarkets are not the key to the
development of this business and others like it.

Another downside to this type of farming is evidenced by the fact that
much of the time, Padraig and Una rely on volunteers from the ranks of
organic enthusiasts to harvest their produce. Labour is a key issue in organics:
as anyone who grows their own vegetables will know, if you don't spray them
with chemicals, they will become lunch for a range of hungry parasites. They
also have to compete for light and air with weeds that have Godzilla-type
strength and persistence. So if you don't spray them you have to weed them,
usually by hand, and prepare to lose some or all of what you grow to pests and
other threats. Therefore, organic crops are smaller, more labour intensive and
more expensive. For small, but growing, businesses such as the Fahys', which
are dependent at present on voluntary labour, government could do one of
two things. They could say that the Fahys run a private enterprise which must
stand or fall on its own, just like everybody else, or they could extend some of
the preferential treatment which has been a feature of the way we do business
with multinationals. Might it not be an idea, for instance, to reverse the
winding down of community employment schemes and extend them to
valuable local enterprises like this one? After all, we are firmly now in the era
where the growth of small indigenous businesses and job creation is the name
of the game.

In terms of the Irish farming picture, organics is still a minor player. Most
growers, particularly in the organic vegetable sector, are small-scale farmers
with small acreages. What makes the heartache and labour worthwhile for

organic producers is often the twenty to thirty per cent premium they are earning over conventional vegetable produce. That's not to say it's this profitable all the time, and on top of that, organic beef, pork, vegetables and chicken all suffer varying market conditions and price vagaries. But it's a growing market and one many farmers in conventional systems are looking at as a way to develop in the future. On the other hand, recession will or has already forced many families to cut back on what they spend on food, and one would think organics would be one of the first luxuries to go. The next few years will be a critical time to see if its long-term future is viable.

IF FARMING IS THAT IMPORTANT, WHY DON'T WE CARE ABOUT IT?

If potential avenues for development of Irish farming are there, why is it that consumers don't seem to care much about it?

We don't care about it because generally rural Ireland and farming are on the cultural slide. While on the one hand much of Ireland watches TV food programmes endlessly and yaks on about good mozzarella, they don't know where the carrot in their shopping basket came from, or what that might mean: if it was shipped in from the Netherlands rather than grown in County Carlow, what are the ramifications of that? Chef Richard Corrigan summed it up to us with characteristic bluntness: 'The Irish know lots about designing kitchens, marble tops to put on their counters and so on, but they know precious little about food.'

The sad fact is that most consumers don't care that Ireland has already lost its sugar beet industry and that our sheep farmers and fishermen are now under serious threat. We all want cheap food, including farming families who go to the multiple supermarkets for their weekly shop like everyone else. The problem is that not only is cheap food often not very good for us, it also hurts farmers. The question is, how many of us see this direct connection or care enough about farming to change our attitude to the food that we put into our mouths and where it comes from? Do we care enough to buy more Irish food? If we as a nation are increasingly disconnected from rural life and from growing food or knowing about food, it's likely that the short answer is that we don't and won't.

The quality of Irish food may be high, but consumer agencies are encouraging everyone to look for the cheapest deal possible. Therefore we have the paradoxical situation where the success of Irish farming as an employer and generator of income is substantially dependent on how badly

Irish consumers want that industry to survive and thrive, and how much they are prepared to personally invest in thinking about it in terms of their weekly shop.

So, if we Irish ourselves are a key factor in keeping Irish farming and the vibrancy of rural Ireland afloat, how many of us know or care enough about the relevant issues to invest in the sector through our own behaviour, specifically in terms of what food we buy? Ten years ago, it wasn't looking so good. Research carried out in 1996 showed that the public had a poor perception of farming and agriculture in general, its importance to the economy, animal welfare practices and care for the environment. Sixty-two per cent of people polled thought that the future of farming and agriculture in Ireland was poor to fair, and thirty-five per cent thought that agriculture was neither important nor unimportant to the Irish economy.[5] So low was Irish people's engagement with agriculture that the Agricultural Awareness Trust set up shop to focus on educating the Paddies about farming.

Strange to think it was even necessary, but there you go. Not surprisingly, their key patrons are the big players in the Irish food sector: Dairygold, Glanbia, Dawn Meats, Kerry Foods and various other big business interests. Initiatives in primary and secondary schools have made up much of their attack, with projects on food awareness such as Meet the Spuds, which in 2008 was generally viewed as an enormous success in creating awareness that— shock horror—much of what we eat grows in dirty brown soil and gets rained on. Following their Healthy Eating Challenge in schools, ninety per cent of students stated that they were more aware of the range of healthy products that can be sourced locally and seventy-five per cent said they would proactively encourage their parents to purchase local produce. That was before they all headed to McDonald's, which despite its image nevertheless buys Irish beef.

Ten years into the job, Agri Aware commissioned new research in 2007 to see whether we were any more knowledgeable about food or buying Irish. Their poll 'shows marked improvements in the public perception of the farming and wider agri community. There have been improvements in the public perception of the importance of the industry to the economy, the contributions farmers make to the countryside and the animal welfare standards employed at farm level.'[6]

Agri Aware has also teamed up with the Association of Home Economics Teachers and the Agricultural Science Teachers' Association, organised healthy eating demos, study guides and school visits by chef Neven Maguire, and its campaigns have won positive reaction from both students and staff involved in the initiatives, with ninety per cent of students more aware of healthy products and how they could be sourced locally. While Agri Aware doesn't claim credit for the changes in perceptions over the ten-year period, it must be acknowledged that the organisation has played some kind of positive role. One of their livelier initiatives is a mobile farm; a collection of livestock that tours schools in order for children to have contact with real farm livestock. You might imagine that providing a sweet cuddly baby sheep for small hands to pet would turn any child off eating lamb, but apparently it has the opposite effect.

There certainly seems to have been a significant improvement in the public perception of the farmer's role, care for livestock and the environment. But how does this translate into our consumer behaviour and our connectivity, or lack of it, to rural Ireland? One thing that has hugely helped attitudes towards Irish farming is the trend towards sourcing local and traceable produce that is happening across the developed world. As a backlash to the highly commodified and packaged food giants, everywhere we look, we are told by chefs and food writers to source local, fresh produce. It's a food fashion that benefits farmers, particularly those in the specialist sectors of the food industry: cheese makers, ham curers, preserve manufacturers and so on. It's all part of the current vogue for cupcake making, nesting and buying courgette seeds you never plant. But this fashion for farmy, cutesy food could prove to be short-lived in relative terms, or too expensive for consumers to cope with. While your artisan-made Parma ham might seem a far cry from the greedy clamour of the commodities market, the animal that it came from ate a small mountain of grain in order to transform from piglet to plump side of pork. Oil and commodity prices will affect all foods, and there's no point thinking that the speciality foods sector is in any way insulated from economic realities; in fact, it is probably more vulnerable.

Bord Bia thinks the sector will survive but that it has, to put it mildly, challenges ahead. What we do know is that the total value of the food we export from Ireland was a hefty €8.2 billion in 2008.[7] Sadly, though, the government doesn't seem too convinced. In the Department of Finance, where they now seem to have trouble calculating their bus fare correctly, agriculture is reckoned to be worth a meagre one per cent of GDP. Little point in trying to develop that, then, when there are brand new shiny toys like 'the knowledge economy' for politicians and civil servants to play with. Perhaps. Except Teagasc, the repository of all agriculture-related knowledge in the state, figures that when you pull in all the related industries, agriculture represents more like eleven per cent of GDP.

Leaving aside yet another example of the left and right hands of Irish government not knowing what they are doing, doesn't eleven per cent sound like a pretty good starting point for something new? Well, not *new* so much as another *Back to the Future* argument: revisiting what we are already good at and looking at how to improve it. There is no need for us to reinvent the wheel in Ireland, but it will take leadership for someone to suggest that. There is a tendency to think that solutions are only solutions when they haven't been tried before. We left agriculture behind us because it seemed part of a past that we didn't want to live. We were all Patrick Kavanaghs, fleeing our own miserable patch of Monaghan. Perhaps now that we have grown up a bit, we can look at our rural roots without prejudice.

CONCLUSION

Too lazy to read it all? Have a look at this instead

If you have neither the time nor the inclination to read this book from cover to cover, this is the chapter for you.

When authors try to convince the public to buy their book, they are sent out to promote it. This usually involves doing interviews with a lot of sceptical print journalists who think they could have done it better and cranky radio presenters who haven't bothered to read it. The whole business isn't an entirely happy one for either party. After several months of spending a dangerous amount of time with their own thoughts, authors never feel they have gotten a fair hearing. The journalists invariably lose patience with the author's inability to sum it all up in less than four minutes. Having played both sides of the pitch on this particular game, we hope that the following hypothetical interview will save everybody a lot of time and effort.

Who should read this book?
Anybody who lives in Ireland and eats.

So who the hell are the two of you to be telling us what's gone wrong with farming and what we should be eating?
Perhaps it's easiest to answer that one by telling you what we are not. We are not a vested interest. We did not come to writing this book with a restaurant, a supermarket, a failing agricultural policy or a hard luck story to tell. The book is an honest assessment after having spent many years standing in rainy fields talking to farmers, looking at food labels, eating our way around the world and writing and broadcasting around these issues.

Are you telling us to do what you guys do? And do you practise what you preach yourselves?
The book is not some kind of manifesto or a 'how to' manual. While we

endeavour to eat good food and support Irish producers, we frequently shop in dodgy places, buy the wrong things and make life easy for ourselves by indulging in Indian takeaways. Yes, we've also bought bagged salads that have been grown in Israel and eat breakfast rolls bought in petrol stations. We are sinners, and probably the very worst kind, because we know we are sinning. We don't want to preach about how to put things right, we're simply trying to analyse why things are the way they are and renew the debate about what we eat and our relationship with where it comes from.

But you've got it wrong from the start when you say that in Ireland we don't love the land any more. There's no nation on earth as mad about land as the Irish.

True, where property and land *ownership* is concerned we are still a bit in love with the stuff. But our relationship with land as a place where food is produced has changed completely. We have fled the land and become an urbanised and, more commonly, suburbanised society, a post-agricultural country. Only the famine and Cromwell driving us 'to hell or to Connacht' have seen greater movements of population. Having moved off the land, our relationship with it changed, and that changed us too. We wouldn't recognise ourselves from fifty years ago, and if we did we probably wouldn't have too much in common to talk about. We have lost our inner culchie.

What is our inner culchie, and doesn't it sound like we are better off without it?

There's no doubt that we have broader, richer lives now. Shifting to a service economy from an agricultural one has freed us from much of the drudgery of life on the farm. And yet we weren't entirely happy being Flash Paddies either. For all the second and third holidays we took a year, the more affluent we got, the more loudly we heaved a sigh of wistful nostalgia. Our rural past was no utopia, but there were things about it that we miss and seek to retrieve. Why else would those who can afford it withdraw back to the countryside and construct pseudo-rustic lifestyles for themselves? Our inner culchie is ying to the yang of the Flash Paddies we became. Michael O'Leary offsets his Flash Paddy emissions by buying a herd of Aberdeen Angus cattle, and many of us embrace a work-heavy suburban life while hankering after a more simple existence and the benefits of a tighter community 'down the country'.

Why is what we eat so important? Isn't it just fuel?

Unfortunately, no. Food and how we eat says a lot about how a society works and how people interact as social beings. We can choose a highly industrialised model like the United States, where food is outsourced, utilitarian, taste-centred and where few people take time to prepare a meal. The by-products of this system are a lack of basic skills in preparing food, things like 'substitute American cheese' in your sandwich and supersized people.

On the other hand, you've places like France and Italy where families and friends devote hours to eating together and yakking on over a table of food. They also spend a fair bit of time talking about food itself, what they cooked that week, a particularly fragrant or possibly foul prosciutto that they bought in the butcher down the street, etc. As people who grow their own veg or enjoy cooking will admit, the growing, sourcing and preparation of food can really be a pleasure. Supermarkets have made us tired, frazzled and half worried about what we buy there, and somehow we know that aisles of packaged food and cheap takeaways can't be the answer.

Ireland isn't Tuscany, but neither should it be Flint, Michigan. There are scary lessons to be learned from handing over total control of our food to intensive farming and the supermarkets. If we commit to this path we will have to deal with the costs to our health, our wallet, the environment and food security in the future. Or we can keep food simple, shop around and enjoy the stuff more.

This whole healthy eating and getting back to the land thing is a wealthy person's luxury. Who has the time or the cash for these sorts of indulgences?

Exactly! Supermarkets and niche producers put the cost of 'quality' food beyond the reach of many. But in the second half of the book we have explored a few ways in which, with the same amount of time and a lot less money, you can eat healthily. The answers are staring us right in the face and they don't involve converting the front lawn into a cabbage patch. A lot of it is stuff that we already knew but forgot; eggs that we didn't get granny to teach us how to suck.

So there's nothing new in your book, then?

Strangely, in the book we talk about a lot of old knowledge that's now kind of new knowledge because we Flash Paddies somehow managed to forget it in the first place. Feeding ourselves was pretty basic until food got turned into a commodity which has been processed and adapted into something we could be made to spend more on. What you put in your gob was never meant to be that complicated. Unfortunately, many of us haven't remembered much of the domestic wisdom accumulated by previous generations. As we moved off the land and into the towns, we thought of it as superfluous and a bit backwards, so we ditched it and embraced the convenience culture offered by food science and the supermarkets. Now we're beginning to wake up to the fact that Gran might have known best, but godammit, we seem to have forgotten much of what she said.

What's wrong with convenience?

Nothing at all, provided that you are prepared to pay more for it and accept that it is not necessarily good for you. Anything, within reason, that saves you labour has to be a good thing. But are we really as helpless and feeble as the supermarkets would lead us to believe we are? Do we really think that cheese should be one-third more expensive because it has been grated for us? Do we really need Marks & Sparks to peel our oranges for us and then re-wrap them in cellophane? Are we not even a little bit embarrassed by being that pathetic?

Those are bad examples. Look at all the genuinely convenient foods that are available.

We pay a price for that convenience, though. We have forced food to do things that it was never supposed to. We have extended its shelf life and made it look more appetising, but reduced its nutritional content. We have replaced natural ingredients with synthetically produced substitutes solely because they are cheaper. Much of the goodness in food has been taken out of it in order for it to look attractive, smell right and last longer on the shelf. Now the food industry turns around and sells us new ranges of 'superfoods' with everything that they took out put back in again.

Why have you got such a problem with supermarkets?

Not all supermarkets, just some. Time and time again, it has been proven that the supermarkets' one true guiding principle is profit. Their bottom line *is* their bottom line. It's not your wallet or your health or your convenience. It's not the environment, the local economy or farmers and food producers. Their sole purpose in life is to grow bigger and get richer.

Are you a pair of hippies in open-toed sandals who won't eat anything other than organic foods?

Not at all. There are a lot of problems with the organic sector. Just because it's organic doesn't always make it better, either. Much of the organic produce on supermarket shelves in Ireland has been flown or shipped in from overseas, with flavour deteriorating over every mile of that journey and lots of carbon being spewed out in the process. Buying food that's made and farmed in a simple manner and close to home is often more important than rigidly sticking to organics. Take the quality of our milk, which will have much more to do with the quality of the pasture, the care the farmer takes of his herd and how the milk is processed than solely with whether or not the cows were fed on organic grasslands. Organics are not a panacea and they need a lot of careful examination.

That said, there can be no doubt that sixty years of heavy reliance on antibiotics and pesticides in food production has left us with a frightening toxic legacy. We probably wouldn't be talking about superbugs killing people in hospitals if we had farmed differently in the past. Water and soil, too, are finite resources: it would be stupid to keep treating them as if they can take any amount of punishment. Commercial farming has left soils at breaking point where they won't produce more food without a lot of chemical fertilization. Rotating crops and old-fashioned methods of keeping soil producing food for us without artificial aid makes a lot of sense.

Are we facing a real food crisis or are you scaremongers?

In 2008, the government embarked on a study to assess the extent of our food security, so no, we are not the only ones worried about food. Think about it this way. Food is a market just like any other financial market. It is run in the interests of its shareholders and has one guiding principle: profit. Left to do its own thing, the market does it very well; it makes money but at the cost of lurching from one food crisis to another. And of course the chief benefactors are the giant corporations that control our food supply, not us, the consumers. 'Every little helps'? Sure it does! We make them loads of cash and they love it.

Because we've become disconnected from food and how it's grown and produced, we have allowed it to become controlled by private enterprise giants. We can't be naive about what that means to us now or, more worryingly, in the future. And it's not just the headline-grabbing stuff, but the rest of it too. Decreasing nutritional value, increased health risks and wild vagaries of price will be the resulting effects of having one main supplier for all the stuff that goes into your mouth. Handing over control of food has also left us anxious about the stuff. How did food shift from being a source of enjoyment to a source of stress?

Is your solution that the state gets more involved in food production?

Think of the food industry as being more than a little like the pre-credit crunch banking sector. Too much influence is concentrated in too few hands. There are people expressing concerns about the way the food industry is being run, but the industry rejects them as Cassandras. Intuitively we know that there are many reasons for feeling uncomfortable about what we eat, but hey, what the hell, party on!

So our options are either to blithely consume our way into a crisis or to do something now instead. Like increase, not decrease, support to farmers producing quality Irish foods. Start at square one, re-educating ourselves and our children about all aspects of food. Ring-fence food or remove it from WTO negotiations. Food production should be a protected industry; it is too important to allow it to be horse traded for Indian car parts or Australian cotton. Transfer power and resources from non-functioning government bodies to ones, like the Food Safety Authority, that have proved themselves. These are just a few of the things that we will be scrambling to put in place in response to a crisis. Why not, for a change, be prepared instead?

Is your argument really that food and farming could be a cure to our economic woes?

Absolutely. All the clever clog economists are agreed that we need to trade our way out of recession because building houses isn't going to make anybody any money for a long time to come. The one sector of the export economy that continues to do well is agriculture, but it is largely overlooked and unloved in favour of courting foreign multinationals that ultimately leave once a cheaper country beckons. Agriculture will never provide as many jobs in a town as an Intel or Hewlett-Packard, but there are up to 300,000 people employed in the food business in Ireland. Crucially, those are jobs that are not subject to the decisions of a board of directors in Boston or Beijing. Irish food is a world-beating product. There are large numbers of Irish food producers that have great products to sell. If they are given better levels of support, there is enormous potential to exploit our natural resources and expertise in farming and food. But agriculture has suffered from political indifference, as we have spent the last twenty or thirty years 'farming the subsidies' and not supporting the agricultural entrepreneurs. It wouldn't be the first time that a solution to our problems has been staring us in the face but we didn't see it. Food is something we're good at, so let's do what we do best.

NOTES

Chapter 1 What's that terrible smell?

1   Reynolds, John, 'Food crisis plan on the menu', *Irish Independent*, 27 November 2008.
2   Darrell, Bruce (2008), 'Planning for Food Security', Feasta Seminar Series.
3   Morris, Steven, 'Veg seed sales soar as credit crunch bites', *Guardian*, 22 April 2008.
4   Brawn, Derek, quoted in www.finfacts.ie/irishfinancenews/article_10010123.shtml, 18 May 2007.
5   Department of Environment, Food and Rural Affairs (UK), https://statistics.defra.gov.uk/esg/ace/pdf/chapter4-ar06.pdf.
6   Newlands, Willy (2006), *Hobby Farming*. Souvenir Press, London.

Chapter 2 Who do you think you are?

1   Todd Andrews, quoted in Farmar, T. (1991), *Ordinary Lives*. A & A Farmar, Dublin.
2   Foster, R. F. (1988), *Modern Ireland, 1600–1972*. Penguin History, London.
3   Cullen, L. M. (1981), *The Emergence of Modern Ireland, 1600–1900*. Holmes & Meier, New York.
4   Young, Michael (1958), *The Rise of the Meritocracy*. Thames and Hudson, London.
5   De Botton, Alain (2005), *Status Anxiety*. Penguin, London.
6   McGinnity, Frances and Calvert, Emma (2008), 'Yuppie kvetch? Work–life conflict and social class in Western Europe'. ESRI Working Paper 239, Dublin, 14 May.
7   Fitzpatrick, Pat, 'Beyond the Pale', *Irish Independent*, 24 August 2008.

Chapter 3 The way we ate

1   Palmer, Mark, 'Prawn again: return of the 1970s', *Daily Telegraph*, 20 January 2007.
2   The Irish Countrywomen's Association was founded in 1910, and it is still the largest women's organisation in Ireland. Traditionally a meeting place for rural woman, local guilds were a place to learn and share knowledge about crafts and gardening and to meet other women. The organisation presently has 15,000 members.
3   McNabb, P., 'Social Structure' in J. Newman (ed.), *The Limerick Rural Survey 1958–1964*. Muintir na Tire, Tipperary, 193–247.
4   Department of Local Government and Public Health, *Reports* (LGD 1930-34:131, various), Stationery Office, Dublin.
5   David, Elizabeth (1999, new edn.), *Mediterranean Food*. Grub Street, London.
6   From the UK's Food Standards Agency website: www.eatwell.gov.uk/healthydiet/seasonsandcelebrations/howweusedtoeat/1990s/.

7   The Slow Food Movement began in Piedmont, Italy in 1989. In Ireland its best known spokesperson is Darina Allen. It has become an international organisation of 80,000 members in ninety countries who not only care about enjoying and retaining our diverse heritage of regional food and drink, and protecting it from globalisation, but are increasingly aware of the associated environmental issues. www.slowfoodireland.com.

8   Gill, A. A., 'Table talk', *Sunday Times*, 6 July 2008.

9   In conversation with the authors.

10 Department of Health and Children (2008), *Slán 2007 Survey of Lifestyle, Attitudes and Nutrition in Ireland*. Stationery Office', Dublin; www.dohc.ie.

Chapter 4 Who is rural Ireland?

1   In 2005, a farmer shot his neighbour in a dispute arising over land. Kilkenny farmer Michael Kehoe shot thirty-year-old Jim Healy dead before turning the gun on himself in what became a double tragedy for their small community.

2   James Mahoney was an artist living in Cork at the time of the famine. In early 1847 he was asked by the *Illustrated London News* to tour the surrounding countryside and report on what he saw. The resulting articles and illustrations did much to alert the British public to the crisis. Mahoney's sparse drawings have become iconic representations of what happened, particularly in the West of Ireland, in the late 1840s.

3   Nassau Senior, quoted in O'Grada, Cormac (1994), *Ireland: A New Economic History 1780–1939*. Clarendon Press, Oxford.

4   L. M. Cullen, cited in Giblin, T., Kennedy, K. and McHugh, D. (1988), *The Economic Development of Ireland in the Twentieth Century*. Routledge, Oxford.

5   Central Statistics Office, 'Agriculture Land Sales by State, Quarter and Statistic', www.cso.ie/px/pxeirestat/database/eirestat/Agriculture%20Land%20Sales/Agriculture%20L and%20Sales.asp.

6   NUI Maynooth, UCD and Teagasc (2005), *Rural Ireland 2025*, Foresight Perspectives, www.finfacts.ie/biz10/FinalForesightReport.pdf.

7   Macra na Feirme press release, 'Young farmers optimistic about the future of farming', 26 October 2007, www.macra.ie/press/show/491.

8   Central Statistics Office, Census of Agriculture Detailed Results, www.cso.ie/px/pxcoa2000/database/census%20of%20agriculture%202000/census%20of%20agriculture%202000.asp.

9   Teagasc (2008), *National Farm Survey*.

10 Keyes, Dermot, 'Agriculture at a turning point', *Munster Express*, 18 September 2008.

11 Crowley, Caroline, Walsh, Jim and Meredith, David (2008), *Irish Farming at the Millennium: A Census Atlas*, NIRSA, NUI Maynooth.

12 Gay and Lesbian Equality Network press release, 14 March 2006, www.glen.ie/press/Challenges.html.

13 'What does rural Ireland really think?', Ploughing Survey Results, *Irish Farmers' Journal*, 18 October 2008.

14 Walsh, Áine Macken, 'Barriers to Change', Rural Development and Sociology Unit, NERC, Teagasc.

15 Éamon Ó Cuív, quoted in *Irish Farmers' Journal*, 8 November 2008.

16 'Why Old Ireland is going up in smoke', *Irish Independent*, 31 March 2008.

17 'What does rural Ireland really think?', op. cit.

Chapter 5 The urban playground

1 BBC News, 22 September 2002.
2 Kirby, Alex, BBC News Online, 20 September 2002.
3 Webb, Tim, 'Over the hills and far away', the *Independent*, 11 February 2007.
4 Whiriskey, John and McCarthy, Paul (2006), *The Irish Sport Horse*. Teagasc.
5 Bradley, Peter, MP, 'Yes, this is about class war', *Sunday Telegraph*, 21 November 2004.
6 Millward and Brown Ulster (2007), Survey prepared for the League Against Cruel Sports, March.

Chapter 6 The land of heart's desire

1 Interview with Maureen O'Hara, *Larry King Live*, CNN, 2 January 2003.
2 *Report of the Commission on Emigration and Other Population Problems, 1948–1954* (1954). Stationery Office, Dublin.
3 *A Book of Saints and Wonders* (1906), *The Kiltartan History Book* (1909) and *The Kiltartan Wonder Book* (1910). She also produced a number of collections of 'Kiltartanese', Lady Gregory's term for the particular mix of Irish and English spoken in the local village of Kiltartan. She produced several versions of Irish myths, including *Cuchulain of Muirthemne* (1902) and *Gods and Fighting Men* (1904). Yeats wrote in its introduction, 'I think this book is the best that has come out of Ireland in my time.'

Chapter 7 Consuming ourselves

1 Thomas, Dana (2007), *Deluxe: How Luxury Lost Its Lustre*. Allen Lane, London.
2 McDougall, Dan, 'Indian "slave" children found making low-cost clothes destined for Gap', *Observer*, 28 October 2007.
3 McShane, Ian, *Irish Times*, 20 September 2008.
4 James, Oliver (2007), *Affluenza*. Vermilion, London.
5 Knight, India, 'Losing my religion in the aisles of the new cathedral of consumerism', *Sunday Times*, 12 October 2008.
6 www.timesonline.co.uk/tol/comment/faith/article3517050.ece.

Chapter 8 Green is the new black

1 Kaufman, Joanne 'Completely unplugged, fully green', *New York Times*, 17 October 2008.
2 Ibid.
3 Herro, Alana, 'Urban agriculture provides Cubans with food, jobs', *World Watch*, 1 March 2007, www.articlearchives.com.
4 Flood, Alison, 'How publishers plan to keep hope alive', *Guardian*, 8 October 2008.
5 Thies, Clifford F., 'The Paradox of Thrift: RIP', *Cato Journal*, VOL. 16, NO. 1, www.cato.org/pubs/journal/cj16n1-7.html.
6 Central Statistics Office (2007), *Environmental Accounts for Ireland 1997–2005*. CSO, July.
7 Smith, Minister Brendan, speech to the Council of Ministers' Meeting, Brussels, 29 September 2008.
8 Ibid.
9 Kavanagh, J. J. (Chairman, IFA Alternative Land Use Committee), *Irish Farmers' Journal*, 4 October 2008.

Chapter 10 Every little helps

1   Tesco press release, 25 February 2008, www.tesco.ie/corporate_info/default.htm?osad campaign=LB4.

2   Tesco press release, 15 April 2008, www.tesco.ie/about/20080415.html.

3   Competition Authority, Determination of Merger Notification, M/08/022-Tesco PLC/TPF.

4   Porter, Sam and Raistrick, Paul (1998), *The Impact of Out-of-Centre Food Superstores on Local Retail Employment*, National Retail Planning Forum (UK), January.

5   Price correct as at June 2008.

6   Competition Authority press release, 'Competition Authority Report Finds Competition between Grocers Is Limited by the Retail Planning System', 10 September 2008, www.tca.ie/NewsPublications/NewsReleases/NewsReleases.aspx?selected_item=225.

7   Competition Authority (2008), *Grocery Monitor Report* NO. 1.

8   Ibid.

9   Ibid.

10 Keynote Ltd (1997), *Retailing in the UK*, cited in George Monbiot (2000), *Captive State*. Pan Macmillan, London.

11 House of Commons All-Party Parliamentary Small Shops Group (2005), *High Street Britain: 2015*. The Stationery Office, London.

Chapter 11 Sense and sensibility

1   Pollan, M. (2008), *In Defense of Food: An Eater's Manifesto*. Penguin, London.

2   Rozin, P. et al. (1996), 'Lay American conceptions of nutrition: Dose insensitivity, categorical thinking, contagion and the monotonic mind', *Health Psychology*.

3   Bratman, S. (1997), 'Health food junkie', *Yoga Journal*.

4   Chan, J. M. et al. (2001), 'Dairy products, calcium, and prostate cancer risk in the Physicians' Health Study', *American Journal of Nutrition*.

5   Fairfield, K. M. et al. (1997), 'A prospective study of dietary lactose and ovarian cancer', *International Journal of Cancer*.

6   Schwartz, G. G. and Hulka, B. S. (1990), 'Is vitamin D deficiency a risk factor for prostate cancer?' (Hypothesis), *Anticancer Research*.

7   Twenty-five cents was the average differential between Superquinn, Dunnes, Tesco, Premier and Avonmore whole milk and no-fat milk: correct as at November 2008.

8   www.foodinnovate.com.

9   *The Grocer*, 'Restoring faith in functionals', 24 May 2008.

10 Advertising Standards Authority (UK) (1996), *Monthly Report*, NO. 60, May.

11 *Which?*, 'Using probiotics in your diet: Good bacteria?', January 2006, www.which.co.uk/advice/using-probiotics-in-your-diet/good-bacteria/index.jsp.

12 *Irish Times*, 'Warning: May contain guff', 28 July 2008.

13 Advertising Standards Authority (UK) adjudication, 12 March 2008, www.asa.org.uk/asa/adjudications/Public/TF_ADJ_44120.htm.

14 *Irish Times*, 'Packing in the claims', 27 October 2008.

15 Margot Brennan, president of the Irish Nutrition and Dietetics Institute, in conversation with the authors.

Chapter 12 Square pegs into round holes

1   Wysocki, Allen, 'Major trends driving change in the US food system', University of Florida, http://edis.ifas.ufl.edu.pdffiles/rm/rm00100.pdf.

2   Greenery Information Service (2003), *The Value of the UK Salad Vegetable Market: UK Salad Market Report 2003*, London.

3   Serafini, M. et al. (2002), 'This new invention to prolong shelf life', *British Journal of Nutrition*.

4   A 1995 study by the UK's Public Health Laboratory Service found that thirteen per cent of bagged salads were contaminated with E. coli.

5   Mary Ellen Camire, Head of Food Science, University of Maine, quoted in *Progressive Grocer*, May 1998.

6   Promotional literature for the Greefa IQS, www.greefa.nl.

7   Food Safety Authority of Ireland (2008), *Trans Fatty Acid Survey (2007)*. Retail Products, May.

8   National Task Force on Obesity Report, 2005.

Chapter 13 Mythbusting: A tale of two markets

1   Battles, Jan, 'Are we too posh for Lidl?', *Sunday Times* (Irish edition), 30 March 2008.

2   National Consumer Agency (2008), *Grocery Price Survey*, 27 February, www.nca.ie/eng/Research_Zone/Reports/NCA_price_survey_27-02-2008.xls.

3   Amarach Consulting (2008), *National Consumer Agency Survey Findings*, March, www.amarach.com/assets/files/NCA%20Consumer%20Research%20Presentation%20March%202008.pdf.

4   Figures supplied by Aldi at the beginning of 2009.

5   Devlin, Martina, 'Status is no small potatoes when class system comes down to earth', *Irish Independent*, 5 July 2007.

Chapter 14 Where does it all come from?

1   Murphy, Michael, writing in the *Irish Farmers' Journal*, 28 June 2008.

2   Vandore, Emma, Associated Press, 12 August 2007.

3   McDonald, Scott, 'Nearly 53,000 Chinese children sick from milk', Associated Press, 22 September 2008.

4   Pearce, F. (2006), *When the Rivers Run Dry: Water—The Defining Crisis of the Twenty-First Century*, Beacon Press, Boston.

5   Bord Bia (2008), 'The future landscape of global food and drink'. Report, Spring/Summer.

6   UN News Service, 'Rearing cattle produces more greenhouse gases than driving cars', 29 November 2006,
    www.un.org/apps/news/story.asp?NewsID=20772&Cr=global&Cr1=environment.

7   Guangrong, Professor Dong, speaking for the Chinese Academy of Sciences, in Lean, Geoffrey (Environment Editor), 'Ice cap roof of the world turns to desert', *Independent*, 7 May 2006.

8   Doornbosch, R. and Steenblik, R. 'Biofuels: Is the cure worse than the disease?', Organisation for Economic Cooperation and Development, Paris, September 2007,
    http://media.ft.com/cms/fb8b5078-5fdb-11dc-b0fe-0000779fd2ac.pdf.

9   Glauber, Joseph, before the Joint Economic Committee, United States Congress, 1 May 2008,
    www.house.gov/jec/news/2008/May/Testimony%20USDA%20Dr%20%20Glauber-Food%20Prices.pdf.

10 Waste and Resources Action Programme (WRAP), 'Wasted food now costs UK homes £10 billion, new study reveals', 8 May 2008, www.wrap.org.uk/wrap_corporate/news/wasted_food_now.html.

11 Quoted by BBC News, 8 July 2008, www.news.bbc.co.uk/1/hi/uk_politics/7494896.stm.

12 Agence France-Presse, 'South American leaders blast EU farm subsidies', 1 July 2008, http://afp.google.com/article/ALeqM5gggB_wYMcliifgdkOZjHT11SqxWQ.

13 O'Toole, Pat, *Irish Farmers' Journal*, 2 August 2008.

14 Economic and Social Research Institute (2008), *Medium Term Review 2008*. ESRI, Dublin, www.esri.ie/publications/latest_publications/view/index.xml?id=2550.

15 Connolly, L., Kinsella, A., Quinlan, G. and Moran, B. (2007), *The National Farm Survey 2007*. Teagasc, Galway.

16 Payments for conversion to organic farming are €212 per hectare for two years and €106 thereafter, in addition to the basic REPS 4 payments. On a fifty-five-hectare farm, the Organic Scheme payments amount to approximately €8,100 per year, or €40,810 over five years. Source: Teagasc, the Irish Agricultural and Food Development Authority, www.teagasc.ie. For other Irish organic resources for consumers or those looking to convert to organic produce, see the Irish Organic Farmers' and Growers' Association, www.iofga.ie.

Chapter 15 Chickens with page three breasts

1 Patrick Wall, Associate Professor for Public Health, University College Dublin, interview with the authors.

2 Rory O'Brien quoted in an article by Olivia Kelleher, *Irish Times*, 8 December 2008.

3 WHO Food Safety Scientist, Peter Ben Embarek, quoted in Lisa Schlein's radio report, 'China's melamine milk crisis creates crisis of confidence', *Voice of America*, 26 September 2008.

4 Dempsey, M., 'No health risk but huge costs', *Irish Farmers' Journal*, 20 December 2008.

Chapter 16 The rise and rise of the superbug

1 Teagasc (2007), 'An econometric analysis of Irish households' expenditure on food away from home (FAFH)', *Consumer Attitudes and Behaviour*, www.bak.teagasc.ie/ashtown/research/foodmarketing/fm-consumerattitudes.asp.

2 Hannah, C., 'Now cooking becomes a game', *Sunday Star*, 22 June 2008.

3 Teagasc (2002), *Food Choice and Consumer Concerns about Animal Welfare in Ireland*.

4 United States Department of Agriculture, Food Safety and Inspection Service, 'Foodborne Illness and Disease' fact sheets, www.fsis.usda.gov/Fact_Sheets/Salmonella.

5 Bernard Vallat, Director of the World Organisation for Animal Health (OIE), speaking at the conference of the Federation of Veterinarians of Europe (FVE), Vienna, 6 June 2008.

6 Speaking as consultant to the US national Centre for Disease Control and Prevention at the World Health Organization, Geneva, 3 May 2004.

7 O'Scanaill, Peadar, 'Out with the old and in with the new', *Irish Farmers' Journal*, 3 January 2009.

8 Albert Osterhaus, Head of the Department of Virology, Erasmus Medical Centre, Rotterdam.

9 European Food Safety Authority (2006), *Community Report on Zoonoses*, Task Force on Zoonoses' data collection on the analysis of the baseline survey on the prevalence of salmonella in pigs in the EU, 2006–2007, EFSA.

10 Williamson, Elizabeth, 'FDA was aware of dangers to food', *Washington Post*, 23 April 2007.

11 Rogers, Gary W. (2008), editorial report, *Journal of Dairy Science* 91: 1279–81.

12　Teagasc (2002), *Food Choice and Consumer Concerns about Animal Welfare in Ireland.*

Chapter 17 They don't go away, you know

1　Department of Agriculture, Fisheries and Food (2006), *Pesticides Residues Report.*

2　Department of Agriculture, Fisheries and Food, statement to the author, October 2006.

3　Metcalfe, M. et al. (2007), 'The economic importance of organophosphates'. California Department of Food and Agriculture, www.cdfa.ca.gov/aes/docs/study/Organophosphates%20in%20CA%20Agriculture.pdf.

4　Lewis, A. et al. (1997), 'A total system approach to sustainable pest management', *Proceedings of the National Academy of Sciences,* 94.

5　Humphrys, J. (2001), *The Great Food Gamble.* Coronet Books, Philadelphia.

6　Osburn, S. (2001), *Do Pesticides Cause Lymphoma?* Lymphoma Research Foundation (USA), www.lymphomaresearch.org/report.php.

7　Ibid.

8　European Commission, Health and Consumers Directorate General (2008), *Monitoring of Pesticide Residues in Products of Plant Origin in the European Union, Norway, Iceland and Lichtenstein 2006.*

9　Pesticide Action Network UK (2005), *The Alternative Pesticide Residues Report,* www.pan-uk.org/Projects/Food/alternativereport.pdf.

10　European Commission (2000), 'Communication from the Commission on the precautionary principle'. Brussels, February.

11　Department of Agriculture, Fisheries and Food (2006), *Pesticide Residues Report.*

Chapter 18 Could food and farming save us all?

1　Power, Jim (2006), *The Importance of the Agri-Food Sector in the Future Development of the Irish Economy.*

2　Central Statistics Office, *Quarterly Household Survey 2004.*

3　Terra Madre Conference, Waterford, 4 September 2008.

4　Young, Peter, 'Why are you falling for organics?' *Irish Farmers' Journal,* 13 September 2008.

5　Bogue, Dr Pat (2007), *Agri Aware: Ten Years On! Review of the Activities of Agri Aware,* Broadmore Research.

6　Bogue, op. cit.

7　Bord Bia press release, 'Bord Bia targets €10 billion food and drink exports by 2011', press release, 28 January 2009, www.bordbia.ie/eventsnews/press/pages/StrategyLaunch.aspx.

INDEX

Abbey Theatre, 82, 84
Abrahamson, Louis, 86
Actimel, 148
additives in food, 156–62, 206
advertising
  farming advertisements, 79–80
  food, 139–40, 143, 148
affluenza, 96, 98, 100, 112, 232
Agri Aware, 224, 228–9
agrichemical industry, 218–20
agriculture see farming
Agriculture, Dept of, 163, 195, 213–14
AIDS, 205
air travel, 107–8, 126, 155, 202–3
Airtricity, 117
alcohol consumption, 49
Aldi, 47, 112, 134, 165–8, 173
  reasons for success, 165–8
All-Bran, 147
Allen, Darina, 45, 239
Allen, Myrtle, 45
Allen, Rachel, 199
Amarach Consulting, 165
Amazon rainforests, 178
America see United States
Andrews, Todd, 21
animal feed, 188–90, 192–4, 195
animal husbandry, 188–210
apples, 157–9
The Apprentice, 26
Arbutus Lodge, 36
Arcadia group, 93
Argentina, 115–16, 179, 183
Asda, 136
Ashbourne, 124
asylum seekers, 29
Attenborough, David, 216
attitudes to farmers, 223–4, 227–30
Austen, Jane, 140, 150
austerity, 3, 4, 106–8 see also thrift
Australia, 175, 178
avian flu, 202, 204
awareness of food, 227–9

B&BS, 62
baby milk scandal, 176–8, 190–1

bachelor farmers, 53–4, 61
Ballaghaderreen, 211–13
Ballinasloe, 58, 61, 226
Ballymaloe, 46, 198
Bangladesh, 176
Bantry, 120
Banville, John, 6
Beaufort Hunt, 67
Beckett, Samuel, 83
Beechlawn Organic Farm, 226
beef production, and global warming, 114–16,
  178–9
Beef Tribunal, 122
Beeton, Mrs, 39
Belfast, 29
Belgium, 191, 192–4
  Belgian Blue, 206–7
Bellevue House, 23
Benecol, 150
Benedict, Pope, 101, 102
betting, 77
Big Houses, 21–5, 80
biofuels, 116, 179
biosecurity, 204
bioterrorism, 203
bird flu, 202, 204
birth control pill manufacture, 193–4
Black Forest gateau, 35
black pudding, 191
Black Tower, 35
Blair, Tony, 69
Blue Nun, 34–5
Blue Stack Way, 73
blue tongue disease, 204
Blumenthal, Heston, 35
Bono, 94
Bord Bia, 213, 229
Bracken, 86
Brackett, Robert E., 206
Bradley, Peter, 76
brand goods, 90–5, 98
Brawn, Derek, 11
Brazil, 115–16, 191
bread, 156
  in animal feed, 188–9
Brennan, Margaret, 121

Britain
  decline of farming, 68–71
  foot and mouth crisis, 69–70
  hobby farming, 12
  hunting ban, 67–8, 70–1, 75
  shopping mall, 99
  supermarkets, 130, 137
  wartime food survival, 4
  waste of food, 181
brucellosis, 202
BSE, 122, 204, 205
building sector, 59, 224
Bunclody, 188–9, 191
burgers, environmental issues, 177, 178–9
butter making, 38
BWG, 136
Byrne, Gabriel, 86
Byrne, Gay, 2, 79
Byrne, Peter, 170, 171

Cadbury, 146, 205
cancer, 218–19
Caraher, Martin, 199
carbon emissions reduction, 113–17
  carborexics, 106–8
Carll, Elizabeth, 107
Carlow, 169–70, 171–3
Carmen Hernandez, Edgar and Maria del,
    197–8
Cashel Blue, 45
Castlebar, 120
Castro, Fidel, 110
Catholic Church, 101–2
  repressive, 84, 101
  sins of globalisation, 101–2
  social values arbiter, 18, 20, 84, 85, 101–2
  stabilising influence on society, 101
  takes over Big Houses, 23–4
cattle, and global warming, 114–16, 178–9
Caulfield, Una, 133
Cedar Club, 24
celebrity culture, 31–2, 94
  celebrity chefs, 36, 198–9
  opening of supermarkets, 128–9
censorship, 84
Centra, 134, 136, 137, 166
The Chastitute, 84
chemicals and pesticides, 211–20, 235
chicken production, 108, 192, 194–5, 210
  rearing your own, 110, 111
child labour
  Irish farms, 38–9

sweatshop labour, 94–5
children, and food, 148
China, 176
  baby milk scandal, 176–8, 190–1
  desertification, 175, 179
  farming, 176, 179, 183
  manufacturing, 92, 224
chlorine, 157
chocolate food scares, 192, 205
cholesterol, 141–2, 150, 160
CIA, 145
class, 19–21, 25–6
Cleary brothers, 225–6
climate change, 113–17, 175–6, 178–9, 216
The Clinic, 87
Clonakilty, 29, 120–1
Clonmel, 133
clothing, and consumerism, 89–103
Coca-Cola, 146
Cockburn, Don, 79
Coillte, 73
Colohan, Ignatius and Pauline, 58–9, 61
commodity market, 180, 183
Common Agricultural Policy, 69, 159, 181
  review of, 183
community, sense of, 2–3, 15, 19–21, 28–30
commuter belts, 60
complaining, 27, 123
composting, 107
connection with countryside, 3, 6, 16–19, 65,
    76, 122
Connemara, 71–2
construction sector, 59, 224
consumerism, 89–103
  ethical, 94–5, 209–10
  loses fashion appeal, 99–100
  thrift, 103, 111–13
  see also shopping
contraception controversy, 85
convenience food, 3–4, 42–3, 47–8, 127, 151–63,
    181, 234
cookery programmes, 198–200
cooking, changing patterns, 34–51
  meal preparation time, 151, 154, 163, 233
Coolmore, 77
Cork, 33
corn, 179
Corrigan, Richard, 3, 198, 227
cost of food
  demand for cheap food, 3, 43, 48, 51, 174, 186,
    227
  rising cost, 48–9, 51, 110, 116, 174–87

country estates, 21–5
country markets, 132–3 *see also* farmers' markets
countryside *see* farming; rural Ireland
Countryside Agency (UK), 68, 69
Countryside Alliance (UK), 70, 74
Cowell, Simon, 6
Cowen, Brian, 74, 182, 196
Craddock, Fanny, 39
Cretu, Arcadie, 212
Cribs, 26
Cuba, 110–11
culchies, 16–19, 22, 32
    'inner culchie', 232
culture, 2–3, 79–88
    connection with countryside, 3, 6, 16–19, 65, 76, 122
    multiculturalism, 18, 29
    popular culture, 80–8
    rural culture, 3, 14, 16–21, 29, 32–3, 63
    urban culture, 58–9, 80
Cumbria, 69–70

Dairygold, 178, 228
Dalai Lama, 51
dance halls, 62
Danone, 146, 148
Darrell, Bruce, 5
Dashwood, Elinor, 143
David, Elizabeth, 40, 41
Dawn Meats, 201, 205, 228
DDT, 219–20
Dear Frankie, 79
deaths notices, 2
de Botton, Alain, 27
decline of farming, 15, 54–66, 120–2, 185–6
DEFRA, 70
Deseine, Trish, 199
desertification, 175, 179
designer goods, 90–5, 98
de Thame, Rachel, 109
de Valera, Eamon, 81
diabetes, 48
Diaz, Cameron, 104, 106
Dillon, John, 120, 121–2, 123, 124, 125
Dineen, Molly, 70
dioxins, 188–92, 195
disposability, 95
diversification, 62–3
Doha Round, 182–4
Doorley, Tom, 48
Doyle, Roddy, 88

drinking patterns, 49
drought, 175, 178, 179
DSV, 95
Dublin, 32–3, 58, 124
    economic boom, 222
    National Spatial Strategy, 222–3
    rural ties, 19, 124
    social structure (1907), 21
    tractorcade in, 124–6
duck farming, 203
Dunbrody House, 46
Dundon, Kevin, 47, 198
Dunne, Dick, 63
Dunne, Fergus, 135, 136–7
Dunnes Stores, 131, 132, 134, 136, 137, 165, 166, 167, 172

E. coli, 157
E numbers, 161
Eamonn a Chnoic Loop, 73
Ear to the Ground, 44–5, 55, 87, 192, 194, 225
eating out craze, 46–7
economic boom, 3, 14, 25–8, 46, 100, 113, 162, 221–14
    affluenza, 96, 98, 100, 112
economic recession *see* recession
economics, 112–13, 163
economy, importance of farming, 184–5, 224, 227–30
Eden, 86, 87
Edgeworth, Maria, 80
Edgeworthstown House, 24
education, free, 21
eggs, 139, 200, 210
Egypt, 176
electricity, 116–17
Emmerdale Farm, 85
emigration, 57, 81
employment in farming, 224, 236
    organic farming, 226
Endosulfan, 220
energy resources, 5, 102, 116–17
    energy saving, 106–8
    renewable energy, 116–17, 179
Ennis, 29
Enniscorthy, 137–8
environmental issues, 71–2, 104–18, 174–220
Environmental Protection Agency (EPA), 195
equestrian activities, 76–7
ethical shopping, 94–5, 209–10
European Food Safety Authority (EFSA), 204
European Union, 181–4, 219

European Union, *continued*
  agricultural policies, 69, 71–2, 159, 181–4
  Doha Round, 182–4
  Emissions Trading Scheme, 114
exports, 229

Fahy, Padraig and Una, 226
Fair City, 87
Famine, 4, 23, 57
  effect on Irish psyche, 57–8
farmers' markets, 3, 169–73
  threatened by supermarkets, 132–3
farming, 5–6, 14–15, 53–66, 120–7, 174–230,
      235–7
  agricultural courses, 60
  animal husbandry, 188–210
  attitudes to farmers, 223–4, 227–30
  bachelor farmers, 53–4, 61
  biofuels, 116, 179
  Britain, 68–71
  connection with land, 55–8, 65, 80
  decline of, 15, 54–66, 120–2, 185–6
  diversification, 62–3
  drudgery, 108–9
  employment, 224, 226, 236
  eu policies, 69, 71–2, 159, 181–4
  farmers–newcomers relationship, 13, 223
  foot and mouth disease/bse, 69–70, 122
  future of, 64–6, 181–7, 225–30, 236–7
  global warming and, 114–16, 175–6, 178–9
  grants, 62–3, 186
  grants cuts, 59
  growing up on a farm, 36–9
  hobby farms, 10–13, 64–5, 110
  importance to Irish economy, 184–5, 224,
      227–30
  incomes, 120, 121, 185, 224
  inheritance, 85
  media, 87
  media advertisements, 79–80
  organic, 12–13, 225–7
  pesticides, 211–20, 235
  sale of farms, 58
  supermarkets squeeze profits, 54, 165, 186,
      226
  tractorcade protest, 120–6
  'urban farmers', 108–11
  wives of farmers, 53–4, 59–60, 61–2
  workers' pay, 211
  world farming, 174–210
Farm News, 87
Farmweek, 87

Fearnley-Whittingstall, Hugh, 104, 198, 199
Feasta, 5
Fendi, 92
Fianna Fáil, 76
The Field, 84
filmmaking, 80–1, 86
Fine Gael, 76
Fitzgerald, Gerald, 148
Fitzpatrick, Pat, 32–3
flooding, 175
Flora, 150
Flynn, Mary, 211–12
Foley, Val, 194
folklore, 83
food, 3–6, 34–51, 126–7, 139–63, 174–237
  additives, 156–62, 206
  advertising, 139–40, 143, 148
  awareness of, 227–9
  changing consumption patterns, 34–51,
      151–2
  cheap food, 3, 43, 48, 174, 186, 227
  convenience food, 3–4, 42–3, 47–8, 127,
      151–63, 181, 234
  exports, 229
  fast food outlets, 44–5, 151
  'foodies', 46
  food security, 4–5, 180, 235
  frozen food, 42, 43, 47
  functional food, 147–9
  global warming and food production,
      114–16, 175–6, 178–9
  gm food, 206–9
  growing one's own, 3, 4, 5, 48, 49, 50, 108–11
  history of human development, 126–7,
      154–5, 174–9, 215
  Irish produce, recent popularity, 45–6, 166,
      227–8
  junk food, 44–5
  labelling, 139, 145–8, 161, 209–10
  labelling and country of origin, 195
  low-fat, 142–5
  nutrition, 49–50, 139–50, 156–62
  organic see organic food
  packaging, 151, 153, 157, 188
  pesticides, 211–20, 235
  processed food, 139–63, 192, 206
  questions and answers, 231–7
  remoteness from food, 3, 127, 181, 200, 227–8,
      234, 235–6
  rising cost of, 48–9, 51, 110, 116, 174–87
  safety, 39, 160–1, 176–8, 188–220
  small specialist producers, 45–6, 65

waste, 158, 181
world food shortage, 174–6, 180
Food Safety Authority of Ireland, 160, 163, 190, 194, 195–6, 201, 206, 213, 236
Food Technology and Innovation Forum, 146–7
foot and mouth crisis, 122
Britain, 69–70
foreign direct investment, 223
Foster, Roy, 23
French cuisine, 40–1
Frugal Food, 111
Full on Irish, 47
functional food, 147–9
Functional Foods Research Centre, 148
functions, 34
future of farming, 64–6, 181–7, 225–30, 236–7

GAA, 31, 32, 65
Gaelic Revival, 82–3
Gaeltacht, 31, 86
Galvin, Gerry, 35
Galway Blazers, 77
Gambia, 203
Gap, 94
Garage, 86–7
Gardeners' World, 109
Gavin, Diarmuid, 110, 199
Gay and Lesbian Action Midlands (GLAM), 61
Gay Clare, 61
gay people, 61
General Theory of Employment, Interest and Money, 113
Germany, 225
Gill, A.A., 47
Gilligan, John and Geraldine, 6–7
Girotti, Bishop Gianfranco, 101
Givenchy, 92
Glanbia, 178, 228
Gleeson Group, 167–8
Glenisk, 225–6
Glenroe, 86
globalisation, sins of, 101–2
global warming, 113–17, 175–6, 178–9, 216
GM (genetically modified) food, 206–9
Golden-Bannon, Ross, 46, 51
Goodman, Larry, 166
Gormley, John, 76, 115
grain production, 175, 177–80
GM crops, 208
prices, 48, 178, 180, 188
The Great Hunger, 84

Greece, 200
Green, Philip, 93–4
green energy, 116–17, 179
green movement, 104–18
Green Party, 73, 76
Gregory, Lady Augusta, 82–3
Gregory, Tony, 74
Grieve, Guy, 50
grocers, old-fashioned, 165–6
grouse, 71–2
growing one's own food, 3, 4, 5, 48, 49, 50, 108–11
Grow Your Own Veg (book), 111
Grubb family, 45
Guerin, Veronica, 7
Guerrilla Gourmet, 50
Guilbaud, Patrick, 36, 47

Haiti, 176
half-doors, 19
Hamm, Ulrich, 225
Hammond, Davy, 19
happiness, 27–8, 96, 98, 100–1
Haughey, Charles, 36, 77
Hazlitt, Henry, 113
Health, Dept of, 163
healthy eating, 49–50, 139–50, 233
diet-related illnesses, 162
processed foods, 156–62
see also safety of food
Healthy Eating Challenge campaign, 228
hedge funds, 180
Hickey, Tom, 86
hillwalking, 72–3
Hinde, John, 19
Hindmarch, Anya, 106
HIV, 202, 205
hobby farms, 10–13, 64–5, 110
Hogan brothers, 213
holiday homes, 14, 56–7, 223
Holland, 193–4, 214
home entertaining (1970s), 34–5
homogeneity, 17, 19, 30–1, 96
horse industry, 76–7
hotels, 24–5
housing, 13–14, 56–7, 223
hunter-gatherer craze, 50–1
hunting, 73–7
ban in Britain, 67–8, 70–1, 75
Huston, John, 80–1
hydrogenated fats, 156, 160

ICA (Irish Countrywomen's Association), 37, 132
ICARUS (Irish Climate Analysis and Research Unit), 175
identity, 2–3, 16–33
  culchies/townies divide, 16–19
  rural identity, 3, 14, 16–21, 29, 32–3, 63, 79
IFA (Irish Farmers' Association), 59, 182
  IFA Countryside, 74, 76
  tractorcade protest, 120, 121–2
images of Ireland, 19, 25, 79–88
  environmental greenness, 79, 114
  romantic images, 79–83
immigrants, 29
India
  farming, 176, 183, 184
  manufacturing, 94–5, 224
Ingram, Alex, 111
inheritance, 85
In Search of Perfection, 35
internet, 61–2, 155
Irish Climate Analysis and Research Unit (ICARUS), 175
Irish Council Against Bloodsports (ICABS), 74
Irish Dairy Board, 178
Irish Farmers' Association see IFA
Irish Farmers' Journal, 16, 53, 64, 72, 87, 192
Irish Farmers' Organisation, 116
Irish Horse Welfare Trust, 77
Irish identity see identity
Irish Independent, 170
Irish language, 31
Irish Masters of Foxhounds Association, 74
Irish Nutrition and Dietetics Institute, 149
Irish stereotypes, 82
isolation, 54, 60–2, 63–4
Ivory, James, 67

Jacobs, Mark, 105
James, Oliver, 96
Japan, 89–90, 91, 147
Jiabao, Wen, 176
Jolie, Angelina, 32
Jones, Tom, 94
Joyce, James, 83
junk food, 44–5

Karan, Donna, 92
Kavanagh, Patrick, 84
Keane, John B., 55, 84
Keep Ireland Open, 73
Kellogg's, 146, 147

Kendal, Felicity, 108
Kenmare, 32
Kennedy, J.F., 85
Kenny, Pat, 120
Kerry Group, 178, 228
Keynes, John Maynard, 112–13
Kilcoole, 86
Killinaskully, 86, 87
Kiltimagh, 18
King, Lonnie, 202
Kinsale, 32
Kinsella, Sean, 36
Kirchner, Cristina, 183
Klein, Carol, 111
Knight, India, 99
Knightley, Keira, 106
Kyoto Protocol, 114, 115

labelling of food, 139, 145–8, 161, 209–10
  country of origin, 195
Labour Behind the Label, 94
land and ownership, 55–8, 80, 126, 232
  inheritance, 85
  land acts, 24
  land prices, 11, 13–14
  public access to land, 72–3
Landmark, 87
Late Late Show, 19, 56, 79
Lawrence, Felicity, 157
Leader projects, 62–3
Leahy, Terry, 129
leptospirosis, 202
Letterkenny, 133
lettuce
  bagged, 157
  pesticides, 215
Lidl, 47, 112, 131, 134, 164–8
  reasons for success, 164–8
Lie of the Land, 70
Lifford, 120
Limerick, 33, 194
Limerick Rural Survey, 38
Lindane, 219–20
Lisbon Treaty, 72, 76, 182
localism, 102–3
Locke, Josef, 21
Locks, 46
Londis, 136
Lord of the Dance, 82
Lough Rynn estate, 24
low-fat food, 142–5
Lowry, Flann and Mary, 130, 131, 132

luxury goods, 90–5, 98
LVMH (Moët Hennessy–Louis Vuitton), 91, 92

Mace, 136, 137
Macra na Feirme, 60, 62
Madonna, 94
Maguire, Neven, 228
Mahoney, James, 57
Mallow, 133
Malone, Tom, 171, 172
Mandelson, Peter, 182–4
Mangan's, 136
Marks & Spencer, 35, 130, 134, 181
  pre-peeled oranges, 152–4
Mars, 146
Mart and Market, 87
Martin, Chris, 106
Marx, Karl, 155
Masterchef, 198
McCafferty, Nell, 19
McDonagh, Martin, 87
McDonald's, 42, 228
McKay, Donald, 166–7
McKeith, Gillian, 3
McKenna, Clodagh, 3
McManus, J.P., 25
McNabb, P., 38
media, 30–1, 84–8
  cookery programmes, 198–200
  domestic life programmes, 44
  farmers' markets, 170
  farming programmes, 87
  food safety, 190
  healthy eating promotion, 48
  homogeneity of, 30–1
Mediterranean cuisine, 40, 41, 44
Meet the Spuds project, 228
Mengniu, 176
meritocracy, 20, 26–7
Mexico, 197–8
microwaves, 41–3
Middleton, Lord, 23
Miers, Tommi, 50
milk, 143–5, 150
  baby milk scandal, 176–8, 190–1
  organic, 225–6
Millstream Recycling, 188–9, 195
Mint restaurant, 46
Mirabeau, 36
Mitchelstown, 189
monkeypox, 202
Moore Abbey, 24

morality, 20, 26
Moss, Kate, 108
MRLS (maximum residue levels), 218–19
MTV, 30, 32
mud, and soil fertility, 216–17
multiculturalism, 18, 29
multinationals, 51, 146–9, 155, 172, 181, 208, 235
Murphy, Michael, 175
Musgraves, 136
mushroom industry, 211–15
Mustard Seed restaurant, 46

'nanny state' control, 162–3, 210
National Consumer Agency, 163, 165
National Farmers' Union (UK), 68
National Food Centre, 200
National Health and Lifestyle Survey, 49
National Parks and Wildlife Service, 73
National Retail Planning Forum (UK), 130
National Spatial Strategy, 222–3
Natraceutical, 146
Nenagh, 130–3
Nestlé, 146
Netherlands, 193–4, 214
Newlands, Willy, 12
New Zealand, 175
niche foods, 45–6, 65
Nike, 94
Nipah virus, 202
Northern Ireland, identity and community,
  28–9
No Sweat, 94
Not on the Label, 157
nouvelle cuisine, 47
nutrition, 49–50, 139–50
  lack of in processed foods, 156–62

obesity, 3–4, 45, 48, 162
O'Brien, Flann, 83
O'Brien, Rory and Monica, 189
O'Brien's Sandwich Bars, 151
O'Casey, Sean, 82
Ó Cuív, Éamonn, 63
offal, 36
O'Hagan, George, 124
O'Hara, Maureen, 80–1
oil prices, 48, 179
O'Leary, Michael, 11, 126, 232
Oliver, Jamie, 3, 44, 199
O'Malley, Donagh, 21
On Home Ground, 86
oranges, pre-peeled, 152–4

organic food, 225–7, 235
  eggs, 200
  European demand for, 225
  farmers' markets, 170
  labelling, 139
  labour-intensive, 226
  pesticides and, 215
  recession threatens future of, 47, 227
O'Scanaill, Peadar, 203–4
Osterhaus, Albert, 204
OutWest, 61
Oxfam, 97

packaging, 151, 153, 157, 188
Paltrow, Gwyneth, 104, 106
Parker, Sarah Jessica, 94
Parlon, Tom, 125
pasteurisation, 143–4
personal ads, 53
pesticides, 211–20, 235
Philippines, 95
pig farming, 188–94, 195, 197–8, 204, 220
planning, 14, 223
  supermarkets, 132, 133–4, 137–8
plastic bag tax, 105
Playboy of the Western World, 82, 88
Ploughing Championships, 62, 64
Poitín (film), 86
Pollan, Michael, 142
pollution, 105
population increase, 174, 180
pork industry, 188–94, 195, 197–8, 204, 220
post offices, closure, 63
potatoes, 39, 159
  spraying for blight, 207–8
Powerscourt House, 24
Pratt, Maurice, 79
Premier Molasses, 194
processed food, 139–63, 192, 206
Product Red, 94
property market, 57, 221
pubs, rural, 63
Punch, 82

The Quiet Man, 80–1
Quinn, Bob, 86
Quinnsworth, 134

rabies, 202
racing, 77
Radharc, 19
radio, 30–1, 32, 84–8

farming advertisements, 79–80
fm/medium wave service, 63–4
rainforests, 178
Ramsay, Gordon, 50, 198
Rathcoole, 25
Rathfarnham, 25
Reagan, Ronald, 163
recession, 3, 24–5, 112–13, 221, 224
  effect on hotels, 24–5
  effect on shopping, 99, 164, 169
  food security, 4–5, 180
  Keynes' theory, 112–13
  threatens future of organic food, 47, 227
recycling, 105
red grouse, 71–2
regional development, 222–3
regulation, 162–3, 206, 209–10, 236
Reilly, Ciaran, 61
Reiter's syndrome, 201
religion, 21 see also Catholic Church
Remains of the Day, 67
remoteness from food, 3, 127, 181, 200, 227–8,
  234, 235–6
renewable energy, 116–17
REPS scheme, 186
restaurants, 36, 46–7
reusable shopping bags, 105–6
rice farming, 178, 179
The Riordans, 85–6
Ritz-Carlton hotels, 24
Rogers, Gary, 209
Roscoff (restaurant), 36
Rossport campaign, 72
Roundstone, 32
RTÉ, 32
  closes medium wave service, 63–4
  cookery programmes, 198, 199
rural Ireland, 3, 6–25, 32–9, 53–78, 222–4, 227
  amenity value, 72–3
  attractions of country life, 7–8, 11
  concerns of rural people, 64
  control over countryside, 71–4, 77–8
  country estates, 21–5
  culchies, 16–19, 22, 32, 232
  erosion of services, 63
  farming a contribution to, 224
  growing up in, 36–9
  housing, 13–14, 56–7, 223
  images of, 19, 25, 79–88, 114
  Irish connection with countryside, 3, 6,
    16–19, 65, 76, 122
  isolation, 54, 60–2, 63–4

land ownership, 55–8
population changes, 60–1
rural development schemes, 62–3, 73, 222–3
rural identity, 3, 14, 16–21, 29, 32–3, 63, 79
sports and activities, 67–78
urbanites migrating to, 6–14
Russborough House, 22
Russia, 175, 179
Ryan, Charles and James, 169–70, 171
Ryan, Declan and Patsy, 36
Ryan, Dermot, 195
Ryan, Eamon, 116

SACS (Special Areas of Conservation), 71, 72
Saddle Room, 46
safety of food, 39, 160–1, 176–8, 188–220
Sainsbury, 130, 136
Saint Laurent, Yves, 105–6
salad
  bagged, 157
  pesticides, 215
salmonella, 105, 200–1, 204, 205–6
salt, in cooking, 50
Sargent, Trevor, 4–5, 106
Sarkozy, Nicolas, 182–3, 184
SARS, 203, 204
Schumacher, Joel, 7
Second World War, 4, 142, 215
Senior, Nassau, 57
Shannonside Mushrooms, 211–14
Sheep's Head Way, 73
Shell, 179
shopping, 139–73, 174, 187
  consumerism, 89–103
  ethical, 94–5, 209–10
  farmers' markets, 169–73
  loses fashion appeal, 99–100
  old-fashioned grocers, 165–6
  thrift, 103, 111–13
  see also supermarkets
shopping bags, reusable, 105–6
shopping malls, 99
Shortt, Pat, 86
Silver Tassie, 82
Simply Delicious series, 45
Singh, Sheotaj, 95
Skerries, 25
Slieve Blooms, 73
Slow Food Movement, 45
smallholdings, 12, 110
Smith, Brendan, 115
Smith, Delia, 111

snacking, 50
social class, 19–21, 25–6
Soda Stream, 41
soil, 216–17
Spar, 134, 136, 137
specialist producers, 45–6, 65
spinach, 215
Spin FM, 30–1
sports, rural, 67–78
Spurlock, Morgan, 48
Starbucks, 90, 177
state regulation, 162–3, 206, 209–10, 236
status anxiety, 27
stereotypes of Irish, 82
Stewart, Duncan, 104, 199
Stewart, Rod, 94
stress, 27–8
Succession Act, 85
Suck Valley Way, 73
sugar beet industry, 227
supermarkets, 3, 128–68, 174, 234
  attractive-looking food, 156–62, 234
  discount stores, 164–8
  effect on local businesses and community,
    130–7
  growth of, 129–38
  kiss principle, 165–8
  layout and displays, 149, 155–7, 159
  loyalty to, 165
  opening of, 128–9
  out-of-season produce, 49, 235
  out-of-town locations, 131–2, 133, 134
  power of, 43, 54, 136, 235–6
  profit-driven, 234, 235
  range of produce, 4, 43, 131, 134, 145
  retail psychology, 149, 155–61, 166–7
  shopping choices, 139–63
  squeeze farmers' profits, 54, 165, 186, 226
  turnover and profits, 129–30, 161
Superquinn, 134, 136, 166, 169
Supersize Me, 3, 48
SuperValu, 131, 134, 136, 137, 166
sustainability, 14, 111, 117–18
sweatshop labour, 94–5
Swift, Jonathan, 80
swine flu, 197–8
Synan, Maureen, 129
Synge, J.M., 82, 87

An Taisce, 71, 73, 134
Tales of the Unexpected, 34
Tarry Flynn, 84

Tate & Lyle, 146
Teagasc, 185, 229–30
television, 30, 84–8
  cookery programmes, 198–200
  domestic life programmes, 44
  farming advertisements, 79–80
  healthy eating promotion, 48
Tesco, 129–37
  compared to discount stores, 165, 166, 167, 168, 172
  effect on local businesses and community, 130–7
  ready meals, 159–60, 181
  ruthlessness, 135, 136
  turnover and profits, 129–30, 161
TG4, 88
Thatcher, Margaret, 163
Thomas, Dana, 92
Thornton, Kevin, 50, 198
Thornton's, 47
thrift, 103, 111–13
Tipperary Water, 167–8
Titchmarsh, Alan, 109
Tokyo, 89–90, 91
Tolka Row, 85
tomatoes, 159
Torney, Alan, 211–12
Townlands (drama), 87
tractorcade protest, 120–6
trans fats, 156, 160–1
Tullamore, 135, 136–7

Ukraine, 179
United States
  cultural influences, 30–1
  nutrition, 142–3
  salmonella, 205–6
urban culture, 58–9, 80
urban farmers, 108–11
urbanisation, 3, 15, 58–9, 232
  global, 176

vaccination, 204
Vallat, Bernard, 202
Vandeleur, John Scott, 23
Vatican, sins of globalisation, 101–2
Velvaert, Jan, 193–4
Venu, 46
Verkest, 193
Vets on Call, 87
Vietnam, 204
Vintners' Federation of Ireland, 63

Viscofibre, 146
Vivo, 136
Vogue, 93

Waitrose, 112
Wall, Paddy, 188
Wal-Mart, 172
Walsh, Joe, 121
Walshe, Padraig, 168, 184
Ward, Peter, 132
wartime food survival, 4
waste, 158, 181
water, as a resource, 178
Waterford, Spin FM, 30
wave power, 117
Wayne, John, 80–1
wealth, 25–8, 96, 98, 100, 112
Westfield, 99
West Nile virus, 202
wheat, 179, 188 see also grain
wholesalers, 136
Wild Food series, 50
Wild Harvest, 135
Wilkinson, David, 76
Williams, H., 134
wind power, 117
wine, changing consumption patterns, 34–5
women
  domestic drudgery, 126, 154
  marrying farmers, 53–4, 61
  migration to urban areas, 61
  role of farmers' wives, 59–60
  work–life conflict, 28
Woodies, 109
World Health Organization (WHO), 162, 177, 202
World Organisation for Animal Health, 202
World Trade Organization (WTO), 15, 76, 181–4
  Doha Round, 182–4
WRAP (Waste and Resources Action Programme), 181
Wyeth Pharmaceutical, 193–4

Yates, Ivan, 223
Yeats, W.B., 82, 83
yoghurt, 161
  organic, 225–6
  yoghurt drinks, 146, 147–8
Young, Michael, 26

zoonotics, 201–5